W9-ALV-329

HOLLYWOOD SCIENCE

HOLLYWOOD

COLUMBIA UNIVERSITY PRESS NEW YORK

SCIENCE

Movies, Science, and the End of the World

Sidney Perkowitz

Columbia University Press
Publishers Since 1893
New York Chichester, West Sussex

Library of Congress Cataloging-in-Publication Data
Perkowitz, S.
Hollywood science : movies, science, and the end of the world / Sidney Perkowitz.
p. cm.
Includes bibliographical references and index.
ISBN 978-0-231-14280-9 (cloth : acid-free paper)—ISBN 978-0-231-51239-8 (e-book)
1. Science fiction films—United States—History and criticism. 2. Science in motion pictures. I. Title.

PN1995.9.S26P47 2007
791.43'615—dc22

2007025542

∞

Columbia University Press books are printed on permanent and durable acid-free paper.
This book is printed on paper with recycled content.
Printed in the United States of America
c 10 9 8 7 6 5 4 3 2 1

To Sandy and Mike, with love, as always

. . . and to the millions who, like me, love science fiction movies

Contents

Preface: A Personal Note

This is a book about science in the movies, which appears mostly as science fiction, and it comes out of my own life as scientist and science fiction fan. When I grew up in the 1950s, science was moving out of the laboratory and into the real world in the beginnings of the nuclear age, the space age, and the computer age. The growing weight of science also appeared in the form of science fiction—and what exciting fiction it was! That golden era featured wonderful stories by Robert Heinlein, Isaac Asimov, Arthur C. Clarke, and others, in magazines, books, and films, all of which I loved and eagerly gobbled up.

Years later, I became a research physicist. I've always wanted to be a scientist and can't quite say that I became one solely because of my early exposure to science fiction. But whether science fact or science fiction came first, the two have always been linked in my life. Many scientists feel the same. It's true that we groan and complain when a science fiction film shows incorrect science or caricatures of scientists; even so, we like to see science presented to the world. After all, science is our subject, and for many of us, an all-consuming one.

The power of the connection that many scientists feel between real and fictional science doesn't necessarily depend on the scientific merit of the fiction. Although the science in a film needs to be accurate, or a reasonable extension of what we know, science fiction stories convey other things every bit as important. Such tales express some of the compelling reasons why people become scientists: sheer amazed wonder at our existence in a fascinating universe, and curiosity about how that universe works and what we'll find in it, from the worlds inside our own minds and bodies to the farthest galaxies.

These moments of awe and curiosity aren't limited to scientists. We've all felt them as children and, if we're lucky, as adults too. So when I write about science in the movies, I'm paying homage to how science fiction, at its best, can bring us back to that sense of marvels yet to be found and how it can inspire those among us, myself included, who dream about becoming scientists.

Atlanta, Georgia, and Cannon Beach, Oregon — Sidney Perkowitz
2006

HOLLYWOOD SCIENCE

INTRODUCTION

Chapter 1

Looking for Science in the Movies?

Check Out Science Fiction Films First

> [Science fiction] movies are . . . weak just where the science fiction novels . . . are strong—on science. But . . . they can supply something the novels can never provide—sensuous elaboration . . . by means of images and sounds. . . . Science fiction films are not about science. They are about disaster.
>
> —*The Imagination of Disaster*, Susan Sontag

Think back to the last science fiction movie you saw. Was it *Star Wars Episode III: Revenge of the Sith*, where humans, aliens, and droids battle one another with sizzling laser weapons or humming light sabers? Maybe it was *War of the Worlds* with Tom Cruise, where literally bloodthirsty aliens scoop up people into huge war machines yet eventually give in to natural forces. Or was it *The Island*, where an evil scientist clones wealthy people to provide them with living stockpiles of spare body parts?

Maybe it was a slightly older movie you rented from Netflix, like the 1997 release *Dante's Peak*, where a brave geologist tries to save a town from a volcanic eruption, shown in spectacular detail. Possibly you watched a much older film, such as *The Day the Earth Stood Still*. This 1951 film shows technology we still can't match, like an advanced spaceship and an intelligent robot, along with a message humanity still needs to hear: be very careful with nuclear weapons.

Whichever film it was, surely it set your pulse racing as aliens attacked, spaceships blasted off, or planets exploded. If it was set on Earth, the excitement may have come from seeing a future version of our world or from the presentation of a massive threat such as an incoming asteroid. The film surely had something else too, a nugget of science or technology that was important to the story: space travel and laser weaponry in *Star Wars*, alien susceptibility to Earthly germs in *War of the Worlds*, or genetic engineering in *The Island*. It might also have shown some of the implications of science for humanity, as in *The Day the Earth Stood Still*. And along with the science and its outcomes, it likely had one or more scientists as characters, like the geologist in *Dante's Peak*.

Science has a role in these and many other films, but not exactly in a realistic way, since real-life laboratories and the scientists occupying them aren't all that exciting. If you've ever visited an actual lab, you probably found yourself in a windowless room where jeans-clad, average looking people twiddled knobs, adjusted lasers, or peered through microscopes. This is laboratory science in action, such as it is. Things are even quieter and more obscure around theoretical scientists, who use blackboards and computers, not lasers and microscopes.

Since watching science as it's really done rarely produces an adrenaline rush, and Hollywood studios want to make entertaining films that sell tickets, science in the movies is usually morphed into science fiction. Labs seen on movie screens display more blinking lights and spectacular computer displays than in real life. The scientists are better looking, too, and not always confined to the lab; instead, they're energetically doing science out in the world, either saving it or plotting to take it over. You'll also see technology far beyond what we now have, or even could possibly have: not just laser weapons and smart robots but spacecraft that travel faster than light, which the theory of relativity forbids, or time travel, matter transmission, and shortcuts through wormholes. Often, too, these films carry a tinge of terror, showing awful calamities arising from science misused, such as widespread nuclear contamination, or from natural events that science can sometimes predict, such as earthquakes.

Not all films involving science are science fiction. Hollywood has made biographies or "biopics" of well-known scientists, such as Madame Curie and Alfred Kinsey, and a handful of fictionalized versions of real scientific events, like *Fat Man and Little Boy* (1989), about the creation of the atomic bomb. There are science documentaries as well, such as *A Brief History of Time* (1991) and *An Inconvenient Truth* (2006), but these are generally not Hollywood products. None of these categories, however, has ticket sales remotely like those of a science fiction blockbuster.

Considering all these types, science appears in more films than you might think. A search of the popular Internet Movie Database (www.imdb.com) using keywords "science" and "scientist" yields hundreds of movies; a search using "science fiction" shows that studios have produced more than 1,400 films of this type since 1902—more than one a month—that were widely distributed in the United States. (The numbers go up dramatically if the search includes television films and series or films distributed outside the United States.)

It may be a surprise that science fiction filmdom dates back to 1902, but the genre has been around almost as long as the movies. As often happens with advances in technology, different people and inventions contributed to the rise of the motion picture. One big moment came in 1891, when Thomas Edison demonstrated his Kinetoscope, which used film moving past a light to present the illusion of motion to a single viewer. By the mid 1890s, inventors were projecting moving pictures onto a screen for an audience, as demonstrated by Edison and others in the United States, and especially by the brothers Auguste and Louis Lumière in France. Their public presentation in a Paris café in 1895 is considered the birth of cinema.

Although the film industry was off and running, early movies didn't tell stories as we know them. They showed only brief slices of life such as an approaching train or a kiss, which were enough to astound audiences. But in 1899, Georges Méliès, a French stage magician turned moviemaker, became a pioneer of narration when he filmed the fairy tale *Cendrillon* (*Cinderella*). Soon stories on film were reaching widespread audiences; in 1909, some 9,000 movie theaters were operating in the United States, typically showing movies that ran ten or twelve minutes.

Early in this era, in 1902, Méliès presented the first true science fiction motion picture, *Le Voyage dans la Lune* (*Voyage to the Moon*)—a fourteen-minute silent film in glorious black and white (the widespread use of sound and color in movies was still years away). The lure of space travel wasn't new, but the notion lacked any scientific plausibility until Jules Verne wrote *From the Earth to the Moon* in 1865. Méliès's film uses Verne's idea of shooting a spacecraft to the Moon out of a huge cannon. There are obvious difficulties with such a device, but the basic principle—that a projectile moving sufficiently fast would break loose from the Earth's gravity and could reach the Moon—is scientifically sound (the magic speed, called the escape velocity, is 25,000 miles an hour).

So *Voyage to the Moon* contains a nugget of real science. But although Verne's story surrounds this nugget with lengthy scientific lectures, Méliès's film skips the rest of the science in favor of fanciful events. The spacecraft with its complement of daring astronomers crashes into the eye of the Man in the Moon, who winces. The explorers aren't troubled by the Moon's lack of air, nor are they much surprised to encounter an active populace of Selenites (played by acrobats from the Folies Bergère) beneath the lunar surface.

These scenes make the film entertaining, presumably more important to Méliès—and to modern movie makers—than getting the science exactly right. But still there's that central nugget about how to escape the Earth's gravity. This puts the film firmly within the science fiction genre, one of whose many definitions is that it consists of stories that extrapolate known science to ask "what if?" Méliès also introduced other features that have become essential, including special effects. His use of multiple exposures and stop-motion cinematography was a forerunner of today's computer-generated imagery, which stunningly portrays imaginary scenes.

Other early science fiction films presented more than a single chunk of science. In 1927, *Metropolis* came along to show the future, or, at least, a future. This influential film by the famed German director Fritz Lang is set in the year 2026, in a high-tech city of towering skyscrapers. Most striking is a character that starts as a machinelike robot, and ends as an exact duplicate of the woman Maria, what we would today call an android.

These advances were barely in sight in the 1920s, and *Metropolis* gives us a world defined by its futuristic technology, as recurs in modern films like *Blade Runner* (1982) and *Gattaca* (1997). *Metropolis* also portrays interactions among technology, society, and government: again, a theme in today's films that is important in the real world as well. The film introduced one other science fiction feature, the mad scientist. Four years before *Frankenstein* presented the world with the quintessential loony researcher, Dr. Frankenstein, *Metropolis* gave us C. A. Rotwang, who has built the robot to replace the real woman he loved and lost, and whose sanity goes downhill from there.

Metropolis (1927). This visionary film about a future society in the year 2026 features scientist C. A. Rotwang (Rudolf Klein-Rogge). He's made the female robot behind him (Brigitte Helm) to replace the woman he lost to Joh Frederson, Master of Metropolis (Alfred Abel, left). The subplot "boy meets girl, boy loses girl, boy builds girl" motivates Rotwang's madness and enlivens the story. *A Golden Eagle; see chapter* 9.

Source: UFA/The Kobal Collection.

These films are classics of a "premodern" era of science fiction films. As film scholar Vivian Sobchack points out in her influential work *Screening Space: The American Science Fiction Film*, the modern era began in 1950, with *Destination Moon*. It was followed by a torrent of other science fiction movies, which established approaches to Hollywood science that continue today.

Destination Moon describes the first rocket flight to the Moon. Based on a story by Robert Heinlein, its scientific exactness helped promote a new, post–World War II age of space exploration. The rocket, its flight into space, and the Moon are accurately presented and look authentic: Chesley Bonestell, still considered one of the most detailed and exact of astronomical illustrators, was a technical adviser.

Destination Moon differs from Méliès's *Voyage to the Moon* in its fidelity to the science, even to the detriment of the story; critics weren't wowed by its plot and characters. This film shows that a movie can do the science right (in a clever touch, it uses a Woody Woodpecker cartoon to explain the physics of space travel, an idea revived decades later when a cartoon character explains cloning in *Jurassic Park* [1993]) and can influence public views of science; but it also shows that science alone can't make a first-rate film without the cinematic elements of a good story, engrossing characters, and so on.

Destination Moon (1950). In the film that initiated a golden age of science fiction, the first humans to reach another world stand on the Moon's surface. After a Woody Woodpecker cartoon explains how rocket ships work, and industrialists combine their efforts to build one, the film goes on to show the space voyage and the Moon itself as realistically as possible.

Source: George Pal Prods/The Kobal Collection.

Few other films of the time displayed such accuracy, but they did introduce additional enduring themes. *The Thing From Another World* (1951), from a story by science fiction author and editor John W. Campbell, is about alien invasion. It tells of an unearthly creature whose spacecraft has crashed in the Arctic, a big, strong, mobile plant that ferociously seeks human blood. Aliens have figured in film after film ever since, right up to *War of the Worlds* (2005).

The Thing also reflects real concerns of the time through the character of Dr. Carrington. Like Dr. Frankenstein, this cold-blooded scientist puts the search for knowledge above human concerns. In the midst of the Cold War between the United States and the Soviet Union, his portrayal represents a contemporary reaction to the nuclear science that had destroyed Hiroshima and Nagasaki and might destroy the world.

Nuclear fears also appear in other films of the era. In *The Day the Earth Stood Still*, after observing humanity and its governments, an alien visitor issues a chilling warning about the consequences of misusing nuclear weapons, enforced by an all-powerful robot. In *On the Beach* (1959), set some time after a global nuclear exchange that is about to wipe out all humanity, a physicist character symbolizes the scientists' role in this catastrophe. As *On the Beach* focuses on science gone bad, the film presents moral issues that science raises, and does so with almost no scientific detail. It's a disaster film, as well, where the disaster is caused by humanity, not nature.

However, if it's natural disaster you want, *When Worlds Collide* (1951) fits the bill. The "worlds" in the title are the Earth and a rogue star that will smash into the Earth and destroy it. An alert astronomer spreads the word about the approaching cataclysm then devises an audacious plan to carry a few people to safety in a spacecraft. *When Worlds Collide* has been followed by a long list of films, such as *The Day After Tomorrow* (2004), where science wielded by a heroic scientist confronts a catastrophe, shown by means of exceptional special effects.

The science fiction films of the 1950s and 1960s were popular with audiences, but few were highly regarded by critics at the time. Even some that have become cult classics, like *Godzilla* (1954)—the U.S. version of the Japanese-made *Gojira*, about a monster dinosaur awakened by U.S. nuclear testing that goes on a rampage—were low budget "B-movies" lacking top stars and good

The Day the Earth Stood Still (1951). Alien visitor Klaatu (Michael Rennie, rear, in the shiny jumpsuit), ordinary human Helen Benson (Patricia Neal), and the powerful robot Gort (Lock Martin), step out of Klaatu's flying saucer in this early modern science fiction film. Klaatu looks human and speaks impeccable English, but he still can't convince Earth's leaders to be careful with nuclear weapons. *A Golden Eagle; see chapter 9.*

Source: 20th Century Fox/The Kobal Collection.

production values (One of these, *Plan 9 from Outer Space* [1959], has been called the worst movie of all time). Just as science fiction pulp magazines of the 1920s and 1930s were considered less than respectable literature, these films were on the fringes of respectable moviemaking. But their themes carried over to the next wave of science fiction films, initiated in 1968 by Stanley Kubrick's *2001: A Space Odyssey.*

Kubrick's film and other big-budget films following it in the 1970s were more than respectable; they carried science fiction to a new plane, displayed in the innovations that *2001* introduced. The film combines future technology like space travel with mysterious alien manifestations and unprecedented, awe-inspiring visual effects. It creates a feeling of cosmic awareness that makes it a landmark work. Steven Spielberg's *Close Encounters of the Third Kind* (1977) also generates feelings of transcendence through its aliens, who first manifest themselves indirectly but eventually appear. Their meeting with humanity has spiritual overtones that create a sense of wonder and give the film its special power.

Also released in 1977, George Lucas's *Star Wars* (now called *Star Wars Episode IV: A New Hope*), has a different impact through glossy technology:

2001: A Space Odyssey (1968). With its mixture of stunning technology like this space station, hints of alien presences, a mystical sense, and special effects more spectacular than any seen before, Stanley Kubrick's *2001* reached far beyond the science fiction films of the 1950s.

Source: MGM/The Kobal Collection.

Close Encounters of the Third Kind (1977). After a series of powerful but mysterious manifestations by aliens, their glowing, overwhelmingly huge spacecraft finally appears as entranced observers watch, followed by the aliens themselves. Steven Spielberg's wondrous images of alien impact on humanity brought awe and inspiration to science fiction filmdom.

Source: Columbia/The Kobal Collection.

spaceships traveling faster than light, personable robots or droids, light sabers, antigravity, and the Death Star, which can smash a planet. It also includes the most stunning array of aliens yet put on screen. All this is in the service of an age-old story that has appeared in many different guises, such as the classic Hollywood western: good versus evil.

In *Screening Space*, Vivian Sobchack names *Close Encounters* and *Star Wars* as marking a major renaissance in science fiction films. Since their release in 1977, she writes, "the genre's production and popularity have increased so greatly that the films and their cultural significance can no longer be easily ignored or dismissed"—unlike the scant critical attention paid to earlier science fiction films. Today, there's no doubt about the power of Hollywood's science fiction films. The low-budget non-prime-time efforts of the 1950s have been replaced by extravaganzas that portray imagined universes through superb sets and special effects. These movies cost tens or hundreds of millions of dollars to make, but they bring corresponding financial returns. A third of the top fifty all-time highest-grossing movies are science fiction, which means that millions upon millions of people have seen them (*Star Wars Episode I: The Phantom Menace* [1999] leads the science fiction pack with $920 million in global box office sales).

Star Wars (renamed *Star Wars Episode IV: A New Hope*) (1977). Abysmally evil Darth Vader (David Prowse, left, with voice by James Earl Jones) and wise Jedi Master Obi-Wan Kenobi (Alec Guinness) battle it out with light sabers. With scintillating images of sleek technology; human, alien, and robot characters; and a compelling conflict between right and wrong, George Lucas's enormously successful film contributed to a science fiction renaissance along with *Close Encounters of the Third Kind.*

Source: Lucasfilm/20th Century Fox/The Kobal Collection.

Given their huge viewership, it's no surprise that science fiction films are now firmly embedded in popular culture and mythology. Just as we all know Cinderella, Dorothy, and the Wizard of Oz, we know R2-D2, Mr. Spock, and Neo, and the phrases "the Matrix," "May the Force be with you," and even "Klaatu barada nikto." In fact, science fiction words have officially entered the English tongue. According to the *Oxford English Dictionary*, "Godzilla" means "A large or strong example of its type," and "Beam me up, Scotty" is an established catchphrase (actually, in *Star Trek IV: The Voyage Home* [1986], Captain Kirk says "Scotty, beam me up"). The OED also notes that the Star Wars series has made "droid" a synonym for "robot," and that President Ronald Reagan's 1983 antimissile Strategic Defense Initiative is called "Star Wars."

The significance of these films, however, goes beyond their popularity and their entry into everyday conversation. Because they can take viewers outside the world as it is to explore the world as it might be, they can provide special perspectives on ourselves and our society, a quality that increasingly attracts film scholars. Sobchack writes that the science fiction renaissance of the late 1970s was accompanied by "a major shift in the . . . practice of film scholarship . . . film scholars have enriched their understanding and description of the cinema as art and cultural artifact by drawing upon the knowledge and vocabularies of other disciplines." Using these tools, film critics and scholars now recognize and analyze the cinematic importance of science fiction films; for example, *Close Encounters of the Third Kind*, *Star Wars*, and six others have made it onto the list, compiled by the American Film Institute in 1998, of the one hundred best American movies.

In the field of film studies, Sobchack and others have examined the narrative and aesthetic richness of science fiction films and the range of meanings these motion pictures carry beneath the surface. Some writers analyze significant individual films such as *2001*; others explore, for instance, what the portrayal of aliens and artificial beings in such films as *Close Encounters of the Third Kind* and *Blade Runner* tell us about our own humanity, or what dysfunctional societies like that in *Metropolis* reveal about our own social order.

Still other writers relate science fiction films, primarily an American product through their Hollywood roots, to American society and history, as M. Keith Booker does in his 2006 book, *Alternate Americas: Science Fiction Film and American Culture*. And some enormously successful films, such as *Alien* (1979) and its sequels and the *Terminator* series (1984, 1991, and 2003), have inspired vigorous debates among critics who discuss how they reflect our attitudes toward everything from feminism to globalization (readers interested in these and other analyses of science fiction film can find references in the "Further Reading and Viewing" section).

Few film scholars, however, deal directly with a major aspect of science fiction films and science in film, namely, the science itself. Yet much of the cultural weight of these films comes from the science they contain. People who've never studied physics, biology, astronomy, mathematics, or geology are exposed to black holes, chaos theory, mutations, volcanism, genetic engineering, nuclear radiation, robotics, and a dozen other topics through the movies. They also encounter science-related catastrophes that could truly affect them and the world, such as the meltdown of a nuclear reactor or the appearance of a deadly virus.

Like the general cultural manifestations that appear in film, these references to science don't arise in a vacuum; they mirror real science and its effects on society. Films made soon after World War II dealt with the space age and nuclear destruction, important topics of the time, and now contemporary fears of nuclear terrorism appear in *The Sum of All Fears* (2002). When real-life computers became more powerful, along came the artificial intelligence HAL in *2001* and humanoid robots in the *Terminator* films. When new biological knowledge made genetic manipulation possible, *Jurassic Park* and *Gattaca* appeared, and *Outbreak* (1995) reflects real possibilities of biological terrorism and warfare. Both film theorists and Hollywood marketing gurus know, though, that the movies also significantly form our attitudes toward the world. Our reactions to science are partly shaped by what we see on screen. As ideas filter from the scientific world into movies, and from movies into the general consciousness, it's important to ask how well the science is presented.

A science fiction film is not meant to be a scientific lecture; still, these films often go over the edge into a world ruled by laws of Hollywood science, not real science. If you were earning a science degree in that alternate universe, here are some questions you might see on your final exam:

1. If we ever meet aliens, they will:
   a. look just like us, except maybe for funny ears
   b. inevitably be repulsive, hostile, or both
2. Should a huge asteroid collide with the Earth, the most likely place it will hit is:
   a. an ocean or possibly a desert
   b. smack in the middle of Manhattan
3. After a nuclear blast, there would be:
   a. total devastation
   b. total devastation, made even worse by unnatural mutants
4. We can use cloning to:
   a. copy plants and animals, but only after time and effort
   b. copy pets and people, even dead ones, right on the spot

5. We need to be very careful with computers, since:
   a. they never work right anyway
   b. they might band together and decide humanity is unnecessary

If you answered "b" to each question, congratulations, you have your Ph.D. in Hollywood science, but not in real science. (The one exception is the first question, since the jury is still out about possibilities for alien life; but judging from the bulk of science fiction films, there's no doubt which answer Hollywood science considers correct.)

So questions like the following are very much in order: How much faith can we put into the dramatized scientific fact and theory in films? Do the technological wonders displayed on screen have a real basis? Could cataclysms like collision with an asteroid actually happen? Do movies about the effects of science on society influence how people act or predict the future?

We should also ask how movies portray scientists because people form impressions of scientists through the media. Some movie scientists are realistically drawn, but many come in only one of three stereotypical flavors: evil, noble, or nerdy. Such simplistic categorizations can make scientists seem strange, which doesn't help science and society relate to each other. So we need to ask: Do movie portrayals show scientists as they really are? Are scientists truly smart? Are there real examples of mad scientists, or heroic ones?

The following chapters respond to these questions by presenting more than a hundred films, chosen to illustrate definite points about the portrayal of science and scientists in motion pictures. These come mostly from the science fiction genre but include some biopics and documentaries. The films are organized under six important scientific themes that also carry potential dangers for humanity—alien life, catastrophes from space, catastrophes from the Earth's own processes, nuclear science, disease and genetic manipulation, and robots and computers.

We begin with a big topic that has inspired many films and is thrilling to contemplate as we explore the universe beyond our own planet: life on other worlds.

PART I

DANGERS FROM NATURE

Chapter 2

Alien Encounters

Scotty: It sounds like . . . some form of super carrot. . . . An intellectual carrot. The mind boggles.

—*The Thing from Another World* (1951)

Palmer [watching the alien scuttle away as a human head with spider legs]: You gotta be f***ing kidding.

—*The Thing* (1982)

U.S. President Whitmore [the president's response to alien invasion]: Let's nuke the bastards.

—*Independence Day* (1996)

Probably since humans first saw the stars in the night sky, we have wondered about the universe: What's out there? How and when did it begin? How big is it? Will it end, and when? What's our place and purpose in it, if any? And knowing that we live in a huge cosmos with cold, empty spaces between the stars, always there is one last question: Are we alone?

That big question is answered with an equally big "No!" in the many films that employ a classic science fiction theme: human meets alien. Since we have no idea if any kind of life or civilization, primitive or advanced, exists out there, we're free to imagine who—or what—we might meet and to shape movie aliens to reflect our own hopes and fears. But imagination has constraints. Although we don't know how life began on Earth, we do know a lot about its physical and chemical basis on our planet. That sets limits, we believe, on what to expect when we meet aliens, if we ever do. As we learn more about the variety of life on our own planet and about conditions in the rest of the universe, however, those limits become broader and more flexible.

Within those limits, movies have run through a gamut of aliens. Some otherworldly beings look just like us. In *The Day the Earth Stood Still* (1951), there's Klaatu (Michael Rennie), the human-appearing emissary from an advanced civilization who lands in a flying saucer to warn humanity to be careful with nuclear weapons or face the consequences. Not only is he like us outside and inside (only better, with a longer life span and marvelous powers of recovery), he also looks good in both a futuristic, nicely tailored silver jump suit and ordinary Earthling wear. But also beginning in the early 1950s, some aliens were decidedly different. In *The Thing from Another World*—also made during the post–World War II flying-saucer craze, when many observers reported seeing fast-moving saucer-shaped aircraft that some believed came from other worlds—the occupants of an Arctic research station find out just how different an alien could be.

The story begins as scientists from the research station and a military team investigate a nearby crash site, where they can dimly make out a saucerlike craft and the remains of an inhuman creature buried in the ice. The lead scientist, Nobel laureate Dr. Arthur Carrington (Robert Cornthwaite), immediately sees the discovery as the key to the stars.

The team salvages the body in a block of ice and brings it back to base, where things take an unfortunate turn. The ice melts, and far from dead, the being (played by James Arness) rises up and looms over a soldier, who shoots it with no effect. Humanoid in shape and seven feet tall, it escapes outside, where sled dogs viciously attack it. After killing three dogs, it runs off, leaving behind a torn-off hand.

The scientists find the hand to be unusual, to say the least. Instead of animal tissue, nerves, and blood, it sports thorn-like barbs and a green fluid like plant sap. This vegetable-like structure, says Carrington, is the reason bullets don't bother the creature, and he doubts that it can die. Visiting journalist Ned "Scotty" Scott (Douglas Spencer) asks if it's a super carrot. Yes, says Carrington, "A carrot that can construct a ship beyond our terrestrial intelligence . . . and guide it sixty million miles or more through space," implying that it may have come from Mars. On the alien planet, says Carrington, vegetable evolution outdid animal development because it wasn't handicapped by emotion or sex. These beings experience no pain or pleasure, which in Carrington's eyes makes them superior to humans. The scientists make one more ominous discovery: this Thing drinks blood for nourishment.

The Thing later returns, killing two scientists and draining their blood. To stop it, the humans try fire, and then a last resort, electricity, triggering a high voltage booby trap as the creature approaches. Huge electrical arcs leap to its head and hands, and in a scene opposite to the electricity-driven birth of the creature in *Frankenstein*, the Thing sinks down and dissipates to nothing—much like the ending of the Wicked Witch in *The Wizard of Oz*. After it's all over, Scotty broadcasts the story over the radio, announcing that an interplanetary invasion may be brewing, and warning the world to "Watch the skies, everywhere! Keep looking. Keep watching the skies!"

The Thing from Another World (1951). Having survived a spaceship crash, Arctic temperatures, gunshots, hostile sled dogs, and fire, this unwelcome alien guest (James Arness) is finally done in by huge electric arcs. The Thing may look man-like, but it's really a vicious mobile plant. The only way to kill it is to cook it like a big carrot. *A Golden Eagle; see chapter 9.*

Source: RKO/The Kobal Collection.

The vision of unemotional, plant-based alien life reappears in another classic, *Invasion of the Body Snatchers* (1956). Physician Miles Bennell (Kevin McCarthy) encounters a rash of odd cases in his California town—a child who says his mother is not his mother, a grown woman who says her uncle is not her uncle. The uncle looks and sounds fine to Bennell, but something is missing, says the niece; authentic emotion has been replaced by simulated emotion.

A psychiatrist friend pooh-poohs the phenomenon, but Bennell suspects that something real and very strange is happening. Sure enough, his friend

Invasion of the Body Snatchers (1956). In a small California town, physician Miles Bennell (Kevin McCarthy, center) with Becky Driscoll (Dana Wynter, right), and Jack Belicec and his wife, Teddy (King Donovan and Carolyn Jones, left), discover something bizarre: an apparently living body that is Belicec's exact duplicate. Later Bennell finds that pods from space are replacing real people with copies that lack all human feeling.

Source: Allied Artists/The Kobal Collection.

Jack (King Donovan) soon discovers a replica of his own body. (When it shows signs of life, Jack's wife Teddy [Carolyn Jones] screams "It's alive. . . . It's alive," another echo of the creation scene in *Frankenstein.*) Bennell finds that huge plant pods, bigger than watermelons, are growing into copies of people that replace the originals; the psychiatrist who tried to allay Bennell's suspicions is himself a replacement.

These pod people grow and distribute more pods. While the originals sleep, their personhood and memories are put into the replicas, which behave almost like those they replace, but not quite. Bennel is unaware that his girlfriend Becky (Dana Wynter) has been replaced until he kisses her, or it, and senses the lack of human feeling. He realizes that the pods come from seeds drifting through space and turn people into creatures lacking all emotion—no love, desires, ambition, or faith—with only one imperative: to survive. However, the film ends with a glimmer of hope, as Bennell escapes from the town, now full of pod people, he finds authorities who believe his story and spread the alarm.

The 1978 remake of the film, with the same title, follows a similar script, except that instead of a small-town doctor, it features an inspector for the San Francisco Health Department (Donald Sutherland, with Kevin McCarthy in a cameo, still spreading the alarm). This version ends more grimly than the original, suggesting that the pods have utterly won and are turning the human race into lock-step zombies.

In either version of *Body Snatchers,* we never see alien entities. There are only the pods. Whatever mind or consciousness animates these aliens, it clothes itself in human bodies. In *The Thing,* for all that the creature is a big carrot, it's manlike in outline. *The War of the Worlds* (1953) is different. It gives glimpses of beings truly meant to look alien.

The film begins as a narrator (Sir Cedric Hardwicke), following the original 1898 H. G. Wells story that is the basis for the movie, explains that Mars is the home of an intelligent but dying race, doomed by lack of resources. In the entire solar system, only our lush world offers the possibility of survival. The next scenes show what seem to be meteors flashing through our atmosphere and landing, but they're not meteors: they're an armada of cylindrical spaceships. Soon flying ships shaped like devilfish or mantas, each with a ray weapon resembling a striking cobra, emerge from the cylinders and wreak havoc. (This differs from Wells's story, in which the war machines are tall constructions that walk on three legs). The rays destroy everything they touch, from people and buildings to military tanks, guns, and fighter planes.

Nuclear physicist and astrophysicist Clayton Forrester (Gene Barry) happens to be fishing near a cylinder landing site in the Los Angeles area, and he is the first scientist on the scene. He surmises that the invaders come from Mars, which is relatively close to the Earth at the time. With his scientific colleagues, Forrester concludes that as inhabitants of a smallish planet with a tenuous atmosphere, the Martians would find the Earth's stronger gravity and thicker atmosphere hard to deal with.

But this doesn't seem to slow down the invaders, and Clayton flees to an abandoned farmhouse with his companion Sylvia van Buren (Ann Robinson). Unluckily, a Martian spaceship plows right into their hiding place. The aliens search for the two humans, and one puts his hand on Sylvia's shoulder. As she and Clayton struggle to escape, we get a fleeting look at a Martian. It has a blob of a body and a big knobby head with a large vision organ made of three segments colored red, green, and blue. It also has two arms ending in three-fingered hands. The arms are twiglike, suggesting that these beings are physically delicate. The humans get away with a sample of Martian blood, which later analysis shows to be anemic.

Anemic or not, the invaders tear through Earth's military forces and destroy the Eiffel Tower in their worldwide rampage. When they prepare to attack Los Angeles, the Army makes a last-ditch effort and drops an atomic bomb on the Martian base. But the aliens throw up a protective force field and emerge through a huge explosion and mushroom cloud without a scratch, ready to march on Los Angeles.

The only hope now is that the scientists can develop a new weapon, but in the chaotic city filled with fleeing people, they can't reach their laboratory. Clayton and Sylvia become separated as the manta ships wreck Los Angeles in a blaze of destructive rays. Despairing Angelenos gather to pray as the city disintegrates, and Clayton and Sylvia reunite in a church, where Martian ships attack and the roof collapses. To the huddled humans, all seems lost, until a nearby manta ship suddenly grows dark, swerves, and crashes. All over the city the noise of destruction dies away as more ships crash. The humans venture outside and approach the wrecked ship, where a Martian arm sticks out of a hatch. Clayton examines the arm and declares "It's dead!" Science couldn't

The War of the Worlds (1953). Although the Martians invading Earth are relatively fragile because they come from a planet with weaker gravity, the touch of a spindly alien hand still gives librarian Sylvia Van Buren (Ann Robinson) quite a turn. Later in the film, scientists examine and discuss other details of Martian physiology, such as their eye structure.

Source: Paramount/The Kobal Collection.

stop the aliens, but a Higher Power could, as the narrator explains in voiceover while joyous crowds gather: "The Martians had no resistance to the bacteria in our atmosphere to which we have long since become immune . . . the Martians were destroyed and humanity was saved by the littlest things which God in his wisdom had put upon this Earth."

While nobody could call the Martians in *The War of the Worlds* handsome or attractive, their big eyes and absurdly thin arms made them look more Smurf-like than threatening, even adding a smidgen of pathos. The Martian flying machines, not the creatures themselves, are intimidating. Machines still dominate Steven Spielberg's 2005 remake (slightly retitled as *War of the Worlds*), but in a different form, and the aliens, though still fragile, are more horrifying.

The 2005 film also begins with a voiceover, but this time, the invaders aren't necessarily Martians; they may come from much further away. Another difference from the 1953 film, and a return to Wells's story, is that the invaders use war machines that operate on land. These tripods, many stories tall and walking on three long tentacles, have been buried underground on our planet for eons. They lurch into life when their alien crews arrive via

War of the Worlds (2005). This remake of the 1953 version features aliens attacking humanity with fearsome three-legged war machines, as in the original H. G. Wells story. Today's special effects generate aliens that are more convincing than in 1953 (see the previous image), but the 2005 version includes no scientists and little scientific exposition about the nature of the invaders.

Source: Dreamworks/Paramount/The Kobal Collection.

multiple lightning strikes that destroy electrical systems and telecommunications. The machines emerge with earthquake-like violence and wield ray weapons that destroy everyone and everything in their path.

The initial scenes are set in New Jersey (probably in homage to Orson Welles's famous 1938 radio broadcast of the H. G. Wells story, which listeners thought described real aliens landing in New Jersey). Among the fleeing populace are divorced father Ray Ferrier (Tom Cruise) and his children Rachel (Dakota Fanning) and Robbie (Justin Chatwin), trying to reach his ex-wife, Mary Ann (Miranda Otto), the children's mother, in Boston. Along the way, they encounter panicked, violent crowds as civilization breaks down. They see the tripods defeat the military and kill without mercy, including capsizing a ferry full of refugees. Most horribly, they watch as the machines use dangling tentacles to snatch up humans for their blood, and sow new plant life around the countryside, red weeds fertilized with that blood.

Ray and Rachel meet the aliens directly when they hide in a basement with semicrazed survivalist Harlan Ogilvy (Tim Robbins). In scenes that copy the 1953 film, their hiding place is searched, first by a visual sensor at the end of a tentacle, and then by the aliens themselves. These look vicious and devilish and move in a monkeylike crouch. Although their thin limbs still seem flimsy, they're infinitely more fearsome than the Martians in the original film.

All seems hopeless for humanity. But when Ray and Rachel (Robbie is off fighting the invaders) wearily enter Boston, they see that some of the red weed is dead, and that a tripod has fallen over. A moment later, another tripod staggers erratically and crashes to the ground. As soldiers surround the machine, a hatch opens and gallons of blood spill out. Another hatch opens, and as in the 1953 version, we see a three-fingered hand move slightly and fall still, followed by the face of an obviously dead alien. Ray and Rachel reunite with Robbie, Mary Ann, and the children's grandparents (Gene Barry and Ann Robinson, from the original film), and a final voiceover tells us that humanity was saved when the tiniest organisms God put on Earth attacked the aliens. Apparently nothing much has changed from 1898 to 1953 to 2005.

Alien (1979) differs from both versions of *The War of the Worlds* in presenting alien hostility through the power of the creature itself, not through machines. Like *The Thing from Another World*, it portrays a being with fearsome natural abilities. The story begins as the crew of the huge mining spaceship *Nostromo* awakens from long term sleep to find that the ship's artificial intelligence, "Mother," has diverted the *Nostromo* from its course. Instead of heading back to Earth, they're going to investigate a signal of unknown origin that may be a distress call.

Landing on the planet where the signal arises, the crew explores a wrecked alien spacecraft. Its interior is extraordinarily eerie, more animal or organic construction than machine. As the exploration continues, "Mother" determines something ominous: the unknown signal is not an SOS but a warning. Meanwhile Kane (John Hurt) finds what looks like a clutch of eggs deep inside the wreck. One egg suddenly opens and something with tentacles shoots out, clinging to the front of Kane's spacesuit helmet.

Kane's shipmates take him back to the *Nostromo*, where Third Officer Ripley (Sigourney Weaver) reminds them that there should be twenty-four hours of decontamination outside the ship when alien life is involved. But Science Officer Ash (Ian Holm) ignores her and lets the group in.

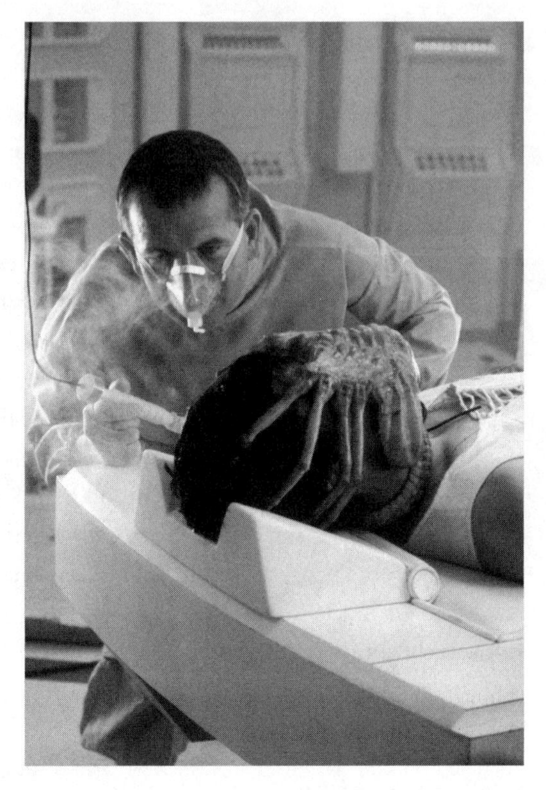

Alien (1979). An alien creature has clamped onto Kane (John Hurt), of the mining spaceship *Nostromo*. As Science Officer Ash (Ian Holm) tries to remove the alien, he finds that it's sheathed in silicon, which in reality has been proposed as an alternative to carbon as a building block for life. Later the creature develops into a huge, horrific, nearly unkillable monster.

Source: 20th Century Fox/ The Kobal Collection.

Kane is laid out in the ship's medical facility with a horror, a spider-crab thing with tentacles, covering his face. He can breathe because the thing has inserted a tube down his throat that feeds him oxygen. Kane might die if the thing is removed, says Ash, but he makes a gingerly incision into the creature. The fluid that comes out, a "molecular acid," instantly burns a hole through several decks. Probing further, Ash declares that the creature has

> an outer layer of protein polysaccharides. He has a funny habit of shedding his cells and replacing them with polarized silicon, which gives him a prolonged resistance to adverse environmental conditions. . . . it's an interesting combination of elements making him a tough little son-of-a-bitch.

Not long after, the thing is found to be gone from Kane's face. When it reappears, the crew kills it and Ripley wants to destroy the carcass, but Ash says it's unique and must be brought back with them. Still, everything seems fine as the *Nostromo* blasts off from the planet to begin the long voyage home, and Kane appears fully recovered. But at a festive dinner before the crew returns to deep sleep, he starts coughing and then goes into convulsions. As the others hold Kane down, something grotesque happens. Blood spurts from his torso, his stomach rips open, and a small but ferocious alien creature emerges. Ash shouts at the others not to touch it, and it scuttles off.

Before the crew can find the alien, it finds them. It has grown hugely, with an enormous elongated predator's head, horrific, slime-dripping teeth, and tentacles. After it kills the ship's captain, Ripley takes command and finds a secret order directing Ash to bring the alien to Earth at all costs, including sacrificing the crew. Ripley confronts Ash and it becomes apparent that the science officer is not human but an android. It's programmed to protect the alien because the corporation that owns the *Nostromo* wants the creature's capabilities. When Ripley asks the android how to kill the alien, it answers, "You can't. [It's a] perfect organism. Its structural perfection is matched only by its hostility . . . I admire its purity. A survivor . . . unclouded by conscience, remorse, or delusions of morality."

The crew tries to escape in a shuttle craft after setting the *Nostromo* to self-destruct, but the alien kills them all except Ripley. In the shuttle, she sees the *Nostromo* explode, and says "I got you, you son of a bitch." Unfortunately, she's wrong: the monster is in the shuttle. Donning a spacesuit, Ripley opens the hatch and knocks the alien into the rocket exhaust. She sends a final report: "Crew . . . dead. Cargo and ship destroyed . . . Ripley, last survivor of the *Nostromo*, signing off," and prepares to enter deep sleep along with the ship's cat, the only other survivor.

Another favorite theme was the idea of alien possession, as expressed in *Invasion of the Body Snatchers*. A different kind of possession appears in the 1982 remake of *The Thing from Another World*, retitled *The Thing* (also *John Carpenter's The Thing*, after the director). The 1951 film was based on John W. Campbell's story "Who Goes There?" but omitted the Thing's ability to adopt the shape of other creatures. In the remake, the Thing is not humanoid but a monstrous tangle of waving tentacles. Even worse news, though, is that it now has the alarming ability to turn into other beings. After surviving a spacecraft crash in the ice, this time in Antarctica, the creature's shape-changing abilities lead to the destruction of a Norwegian research station.

The alien escapes in the form of a sled dog, which is taken in by the unsuspecting staff of a nearby American research station.

The station's inhabitants soon learn that the dog isn't exactly friendly old Fido. Resident scientists find that the creature is made of cells that can absorb those of any other being and adopt their function. They conclude that if the alien escapes from Antarctica, it could take over every person on Earth. That lends urgency as the station's occupants, led by helicopter pilot R. J. MacReady (Kurt Russell), go through an orgy of fear and paranoia as they try to discern who is human, who an alien facsimile. Several horrific encounters with the alien include one in which it grows spider legs on a human head. Finally, the station's inhabitants are pared down to MacReady and Childs (Keith David). MacReady destroys the monster with dynamite, but the explosion burns down the station and we're left only with two people ready to wait out the Antarctic weather, if they can.

The Puppet Masters (1994), from the 1951 Robert Heinlein book of the same name, brought another kind of scientific gloss to the idea of possession. A spaceship has landed in Iowa, which interests a secret federal agency called the Office of Scientific Intelligence, especially since the locals are pretending the landing never happened. After the first OSI agents sent in go missing, a new team—Andrew Nivens (Donald Sutherland), head of OSI; his son Sam (Eric Thal), an OSI agent; and Mary Sefton (Julie Warner), a NASA expert in extraterrestrial biology, or exobiology—arrives and finds strange behavior. Mary, an attractive woman, notices that not one man or teenage boy gives her body the up-and-down leer she usually gets.

When Andrew starts asking questions, a local TV executive tries to shoot him. Sam beats the man to the draw and kills him, but the body seems to keep moving. The motion comes from a slug-like parasite entrenched on the man's back, under his shirt. It resembles a small manta ray, with a mottled hide and a long whiplike appendage it can project at high speed. Sam kills it, and after evading other slug-infested humans, the team takes the dead alien back to OSI headquarters in Washington, D.C. On the way, Mary muses that this is not how she had envisioned first contact with aliens, and Sam asks if she was perhaps expecting E.T.?

Mary examines the alien and finds that the parasites are equipped to take complete control. They're mostly brain, with limited physical capability and no sense organs. The long appendage terminates in a probe that directly enters a host's brain, and the creature hooks itself to the spinal cord as well. The aliens also adapt to the host's physical needs, and boost adrenaline flow in their hosts to make them stronger. But in a nod to the vegetable Thing and the pods in *Body Snatchers*, the slug's internal chemistry is based on chlorophyll, not hemoglobin.

Despite all precautions, though, one of the slugs infects Sam. In an eerie scene, Andrew interrogates his son who responds like one possessed. Asked "What do you want? Why are you here?" the slug answers in Sam's voice, saying that the previous hosts were no longer useful. Humans are stronger, they will last longer. Accessing Sam's memories, the slug tells Andrew he's a domineering and lonely man and adds "You won't be lonely when you are one of us. No one will be." Seeing Sam's agony during all this, Mary kills the slug with a jolt of electricity. But another slug takes over Mary. She tries but fails to reinfect Sam while seducing him and then escapes, still slug-ridden.

Meanwhile, the infestation is spreading from Iowa, where the slugs have taken over the police and the army. The aliens reproduce by division every twelve hours, and the alarming news is that there will be 250 billion of them in two weeks. After Sam learns that they congregate in hives, he parachutes into Des Moines and finds their main hive. It's filled with slugs and humans connected together in a kind of mass mind that cajoles Sam to join, promising an end to loneliness. Instead, Sam spots Mary; but before they escape, she shows him a room festooned with hanging human bodies, except for a young boy who's still alive. As all three flee, Mary comments that the child must somehow have been fatal to his slug.

Back again in Washington, the scientists find that the boy's condition does provide a magic bullet. The child has encephalitis, an inflammation of the brain caused by a virus. It's curable if caught early but is quickly fatal to the slugs with their relatively greater amount of brain matter. The disease can be spread by mosquitoes, and this strategy works to stop the slugs in their tracks without harming their hosts.

That isn't quite the end, though. When Andrew, Sam, and Mary visit the ruined Des Moines hive, the last remaining slug attaches itself to Andrew. He escapes in a helicopter, ready to start a new onslaught of slugs, but Sam swings aboard as the craft lifts off. He kills the slug, though only after shooting his father in the shoulder. When they land, there's a slightly awkward moment as they face the fact that the son shot his father:

> **ANDREW:** I can't believe you shot me.
>
> **SAM:** Well, what would you have done?
>
> **ANDREW (CASUALLY):** Oh, I'd have shot you, of course . . . but I never dreamt you were that kind of man

Somehow, this makes the two bond more closely, and as the film ends, it's clear as well that Sam and Mary will stay together.

Another theme explored in both versions of *War of the Worlds*, mass invasion or armed conflict with aliens, also surfaced in *Mars Attacks!* (1996), *Independence Day* (1996), and *Starship Troopers* (1997). *Mars Attacks!* is a deliberate spoof of early science fiction tales of space invasion that repeats the enduring clichés: hordes of silvery, spinning flying saucers; evil Martians with skull-like heads who kill for the joy of it; ray guns that disintegrate people and things; and earnest speeches to the populace from the clueless U.S. president (Jack Nicholson). One unusual feature of the film is that unlike many movie aliens, the Martians need space suits to survive in Earth's atmosphere.

In *Independence Day*, scientists find that a huge unknown body is approaching the Earth. It can't be a natural object because it's slowing down rather than speeding up. It turns out to be a truly enormous spaceship, nearly 350 miles across and a quarter the mass of the moon. The ship splits into three dozen smaller units, each still stupendous at fifteen miles across and each stationed above a major city.

U.S. president Thomas Whitmore (Bill Pullman) comments that now we know the answer to the question "Are we alone?" Although David Levinson (Jeff Goldblum), an MIT graduate who works for a cable company, begins to analyze patterns in the alien radio signals, all attempts to communicate with the ships fail or are met with hostility.

The alien intentions become unmistakable when one ship hovers over the White House, opens a gigantic maw, and emits an enormous energy beam that blasts the building to atoms. "We're being exterminated," says President Whitmore as the aliens take out Washington, New York, and Los Angeles. Humanity's efforts to fight back are futile, until help comes from an unexpected source: Area 51, the fabled secret government laboratory in Nevada, where Whitmore and his entourage arrive along with other survivors. These include U.S. fighter pilot Captain Steve Hiller (Will Smith), who shot down a small alien aircraft and brings its unconscious pilot with him.

Scientists in Area 51 have been secretly studying an alien spacecraft that crashed in Roswell, New Mexico, in 1949. Now chief scientist Dr. Brackish Okun (Brent Spiner) examines the alien Hiller brought in. It breathes oxygen and can survive in our environment. Although the alien ships seem invincible, the alien itself wears "biomechanical" armor with tentacles, indicating that it's vulnerable and can be killed. The creature walks upright on two legs, has two arms with fingers, and a big head shaped like a devilfish or the tip of an arrow. It has two eyes and also ears, but no vocal cords, suggesting that the beings are telepathic. In fact, when the alien awakens, it almost seizes control of Whitmore's mind but is shot dead first.

However, the brief mind contact showed Whitmore what the aliens want. "They're like locusts," he says, adding, "They're moving from planet to planet . . . their whole civilization. After they've consumed every natural resource they move on . . . and we're next. Nuke 'em. Let's nuke the bastards." But even nuclear weapons can't penetrate the protective shields around the alien ships. Fortunately, computer whiz David Levinson has an idea. Using the captured alien craft, Hiller flies David up to the immense mother ship, where David plugs his laptop computer into the alien system and unleashes a virus that takes down the shields. To finish the job, an international fleet of aircraft attacks the now vulnerable ships. President Whitmore, ex-military, enters the battle in his own fighter jet, but first, since it happens to be July Fourth, he pumps up the assembled pilots with a rousing speech that could have come right out of Shakespeare's *Henry V*:

> You will once again be fighting for our freedom. . . . Not from tyranny, oppression, or persecution . . . but from annihilation. We are fighting for our right to live. . . . And should we win the day, the Fourth of July will no longer be known as an American holiday, but as the day the world declared in one voice: "We will not go quietly into the night! We will not vanish without a fight!" . . . Today we celebrate our Independence Day!

Now human weapons work just fine against the unshielded alien ships. They're shot down as immense piles of wreckage, while the pilots celebrate their victory along with all humanity.

In *Starship Troopers*, humanity faces a different challenge from an alien race. Loosely based on Robert Heinlein's 1959 book of the same name, the film tells of a future war with the Arachnids, or more commonly, the Bugs. Both humans and Bugs have been expanding throughout the galaxy, seeking similar types of planets, which brings them into conflict. The creatures are implacable, not to be reasoned with or even contacted, and have no compunction about slaughtering human settlers on distant planets. Total war begins when the Bugs attack Earth from space, destroying the city of Buenos Aires with a massive projectile (which I discuss in the next chapter).

As humanity fights the Bugs, it learns that they come in a variety of unpleasant forms including "plasma bugs" that emit deadly high-energy beams against spacecraft. The main battle is on the Bug's home planet Klendathu, where the soldiers of humanity's crack Mobile Infantry encounter waves of frenzied Arachnid foot soldiers. These warriors are a hideous combination of spider, scorpion, and praying mantis standing taller than a man. The sickening machinelike motion of their limbs covers ground with incredible speed. Insectlike, they're programmed to fling themselves on the enemy and kill, regardless of losses, until

Starship Troopers (1997). Mobile Infantry trooper Johnny Rico (Caspar Van Dien) is menaced by a giant member of a hostile alien race, the Arachnids. Special effects made it possible to show convincing Arachnids in beetlelike, spiderlike, and flying forms; however, there are fundamental reasons why tiny creatures can't evolve into enormous versions like this one.

Source: Columbia Tristar/The Kobal Collection.

they overrun their opponents. Taking advantage of this hive mentality, the humans find and capture a "brain bug" that controls the Arachnids. This proves to be a first step toward human victory, the beginning of the end of the war.

Not all film aliens are inherently hostile. *E.T.: The Extra-Terrestrial* (1982) presents a vision of friendship growing between a human child and an alien stranded on Earth, who overcome their initial mutual strangeness. *Close Encounters of the Third Kind* shows aliens as ultimately benign and bringing spiritual benefits to humanity. In the *Star Trek* series that began with *Star Trek: The Motion Picture* (1979), the crew of the starship *Enterprise* battles hostile Klingon aliens but is itself a mixture of humans and beings from other planets, including Mr. Spock from Vulcan and the Klingon warrior Worf, working together mostly in harmony. And beginning with the famous cantina scene in *Star Wars* (1977), that series has featured a variety of aliens, some very strange, that inhabit the galaxy with humans and work side by side with them against the evil empire. This includes the huge and bearlike Chewbacca, a Wookie; the minuscule and wise Jedi master Yoda, with his green skin and pointed ears; the tribe of cute and cuddly Ewoks; and others.

From fragile Martians in their machines to Bug warriors that can rip a man apart with the sweep of a claw, from a ferocious alien on the *Nostromo* to Mr. Spock's cool logic in *Star Trek*, from swarms of slugs in *Puppet Masters* to happily dancing Ewoks in the Star Wars series, movies have presented a huge variety of aliens and their works. No matter how convincingly these creatures are presented in the movies, though, we should remember that they're purely imaginary. At the moment, we've not found life anywhere other than on Earth, and scientists don't know enough about the universe to predict with any certainty whether we will. There is, as yet, no established science of aliens; there is, however, solid science that supports one remarkable result, which is that our odds of actually encountering alien life have increased hugely since these films were made.

In the 1950s, the only planets we knew were the nine in our solar system—in order from the Sun, Mercury, Venus, Earth, Mars, Jupiter, Saturn, Uranus, Neptune, and Pluto. (In 2006, the International Astronomical Union reclassified Pluto as a "dwarf planet," leaving the solar system with only eight true planets. Like a planet, a dwarf planet is round and orbits a sun but has not swept the area around its orbit clear of the smaller bodies that arise as planets form. This new category has been controversial among astronomers, and there might be more changes in the future.) We made educated guesses about which of these planets might support life. All earthly life uses a common building block, the element carbon. A carbon atom has four chemical bonds that enable it to hook on to a variety of other atoms, especially nitrogen, oxygen, and hydrogen, to make organic compounds. These are important because they can form intricate molecules like DNA, integral to earthly life.

One of the best media for these molecules to form is water, a second necessity for life as we understand it, since compounds dissolve and recombine into complex new forms in liquid water; this implies temperatures not too hot or cold, where water is neither vapor nor ice. Another requirement is an energy source, like sunlight. A planetary atmosphere, likely containing oxygen like our own, is also important. Reactions with oxygen provide energy, and ozone—a form of atmospheric oxygen—blocks overly intense ultraviolet radiation from the sun. Ultraviolet light can provide energy, but too much of it can disrupt complex biomolecules.

Within these guidelines, pickings were slim. Mercury is barren, with no water or air, and lay too close to the Sun. The huge worlds Jupiter, Saturn, Uranus, and Neptune are distant from the Sun, and since these so-called gas giants are made mostly of hydrogen and helium, offer no solid surfaces. Pluto is another barren body, icy and extremely cold. Only our two neighbors in space had possibilities. Venus is almost Earth's twin in size and has an atmosphere, but the planet's perpetual cloud cover made it hard to determine more. Mars was the favorite. Smaller than Earth, it possesses an atmosphere, although a thin one. It looks desertlike, but its polar icecaps offered the possibility of water. And although it is colder than Earth, the temperatures are less extreme than on the outer planets.

Now, thanks to explorations with spacecraft and planetary landers, we know a lot more about the planets. We also know more about how life can express itself under extreme conditions. Putting all this together, the odds for past or present life have become better in our own backyard, and, as we'll see, in the rest of the universe, too.

In our solar system, it's true that nothing we've learned about the four outermost planets, the dwarf planet Pluto, or Mercury makes them any more suitable to harbor life. Also, having peered beneath the clouds of Venus, we now know it's a hellhole, with surface temperatures near 900 degrees Fahrenheit. Its dense atmosphere is filled with droplets of sulfuric acid and exerts crushingly high pressure. But Mars has new possibilities, and as a remarkable bonus, we've also found that some moons in the solar system are in the running to support life. Another bonus was the discovery, in July 2005, of a new object bigger than Pluto, and orbiting the sun far beyond it. This body, once informally called Xena after the hero of the television series *Xena: Warrior Princess*, is now officially named Eris, and it's the largest dwarf planet we know. Its distance from the sun's warmth makes it unlikely to support life, but the discovery shows that our own solar system still has surprises.

In our immediate neighborhood, astronomers have long aimed their telescopes at Mars. In the nineteenth century, Italian astronomer Giovanni Schiaparelli saw something startling—long, dark, straight lines crisscrossing the Martian surface. He called them *canali* or "channels" in Italian, which became mistranslated as "canals," implying that they were artificial. Later the American astronomer Percival Lowell became caught up in the idea and observed many more supposed canals. By the early twentieth century, a mythology had sprung up in which a sophisticated Martian civilization survived by building an immense system to distribute water from the polar ice caps. Naturally, the inhabitants of this deprived desert world would jump at the chance to take over a lush planet like ours, the basis of H. G. Wells's *The War of the Worlds*.

The canals have proven to be illusory, and modern close-ups of Mars show no signs of civilization. Still, the search for Martian life continues. In 1976, NASA's *Viking* lander ran tests on Martian soil that generated considerable excitement. The results seemed to indicate biological activity, but this turned out to come from nonliving chemical reactions. In 1996, another startling report came from scientists who analyzed a Martian meteorite, a piece of the planet that had fetched up in Antarctica. The meteorite contained what seemed to be ancient fossilized bacteria and other signs that microbes once thrived on Mars, but further study has made these conclusions very uncertain.

However, there is strong evidence that Mars was once wetter than it is now. That idea is supported by the observation that some of the planet's surface features resemble dry riverbeds or lakes, by other geological evidence, and especially by results garnered on the surface of Mars in 2004 by NASA's *Opportunity* planetary rover. This robot analyzed rocks that look as if they were once immersed in salt water, such as what would be found in a shallow sea. So although there's no direct evidence of life on Mars, we are nearly certain that an appropriate environment once existed there—not in the recent past, but billions of years ago. (Also noteworthy: in early 2007, Jeffrey Plaut at NASA's Jet Propulsion Laboratory and his colleagues announced that radar data from a spacecraft orbiting Mars showed enough frozen water at the planet's south pole to cover all of Mars to an approximate depth of thirty-six feet if it were liquid.)

Other candidates to support life are not planets but satellites. Between them, Jupiter and Saturn have nearly sixty moons, some quite substantial. The two biggest, Jupiter's Ganymede and Saturn's Titan, are each larger than Mercury and Pluto. There is remarkable diversity among these bodies. For instance, Jupiter's moon Io has active volcanoes, and Titan is the sole satellite with a dense atmosphere.

The fascination with these moons is that some may contain liquid water, and one, Titan, has a rich mixture of potential building blocks for life. The leading candidate for literally oceans of water is Jupiter's moon Europa, about the size of our own moon. Images from NASA's *Galileo* spacecraft show that its surface resembles blocks of ice floating on a liquid sea. Europa's surface is more than cold enough to freeze water, but its interior could be warm because of what are called gravitational tidal forces. As Europa orbits Jupiter, the changing gravitational pull flexes the moon like a ball of clay being kneaded between two hands. This produces heat deep within the body, which may be enough to keep whole oceans from freezing below miles of surface ice.

Very recently, even more dramatic evidence of internal reservoirs of water came from NASA's 2004–2005 *Cassini-Huygens* space mission to Saturn and its moons (in the seventeenth century, Dutch astronomer Christiaan Huygens discovered Saturn's rings and Titan, and Italian astronomer Jean-Dominique Cassini found other moons of Saturn and more about the rings). In 2006, scientists reported that the mission had obtained images of huge geysers of ice and water vapor emerging from Saturn's moon Enceladus. These are thought to come from pools of liquid water inside the moon, which are kept from freezing by gravitational flexing.

The *Cassini-Huygens* probes also confirmed another unusual feature on Titan. This moon's hazy, orange-colored atmosphere is denser than ours, and like ours, is mostly nitrogen. The color comes from a rich stew of "prebiotic" organic compounds that could eventually form amino acids and other substances essential for life. This atmosphere seems to resemble the atmosphere that surrounded the Earth billions of years ago, before planetary processes and the emergence of life turned it into the oxygen-rich mixture we breathe today. Examining Titan may help us understand that complex evolution, or maybe life is evolving there right now. And leapfrogging the scientific reality, Titan has already played a role in science fiction as the home world of the slugs in Robert Heinlein's story *The Puppet Masters.*

The other big news is that the search for life is no longer confined to our solar system, for we've learned that its handful of worlds aren't alone in the universe. Hints of planets circling other stars came in the late 1980s, and a definitive observation came in 1995. These have continued, and now we know of more than 200 extrasolar planets orbiting stars that lie light-years away.

These worlds are generally too distant to see in telescopes. Instead, we detect them by indirect methods, such as observing their gravitational effects. If a star has an orbiting planet, the star wobbles in space, just as you would wobble slightly if you swung a heavy brick in circles around your body. Measuring the wobble gives the mass of the planet and its distance from the star. Sometimes it also happens that the planet periodically passes in front of or "transits" its star along our line of sight, partially blocking the star's light. When this happens, astronomers can measure the planet's diameter and something about its composition as well.

Most of the new extrasolar planets seem to be gas giants resembling Jupiter in size, but they lie closer to their suns than does Jupiter. These "hot Jupiters" offer unusual features. One, tentatively dubbed Osiris, orbits a star 150 light-years from Earth. From information gathered as it transits its star, astronomers have determined that its atmosphere contains hydrogen, like Jupiter, but intriguingly, also carbon and oxygen.

Although the extrasolar gas giants are different from Jupiter, they still don't seem like prime candidates to support the kind of life we know, but there are more likely possibilities out there. In 2005 astronomers discovered a much smaller planet circling the star Gliese 876, fifteen light-years from Earth. The planet, Gliese 876d, is innermost in a mini–solar system of three planets. It's only seven times as massive as the Earth, whereas a gas giant like Jupiter is 300 times as massive. This suggests, although it doesn't prove, that the new planet has a rocky composition like the Earth's.

Still, Gliese 876d seems less than hospitable to life. It orbits very near its sun, making its surface extremely hot and exposing it to damaging radiation. But in

April 2007, astronomers reported the best bet yet, a planet only five times as massive as the Earth circling the star Gliese 581 in the constellation Libra (the Gliese catalog lists all known stars within eighty light-years of Earth). This world, Gliese 581c, also orbits near its sun, but there's a twist: Gliese 581 is relatively dim, much less powerful than our own sun. Initially, Gliese 581c was thought to lie in the "habitable zone," with surface temperatures that sustain liquid water. Now it appears that its atmospheric conditions make it too hot, but its sister planet Gliese 581d lies further out, perhaps making its temperature just right.

The increasing pace of discoveries of extrasolar planets such as Gliese 581d and Gliese 876d enormously increases the chances of finding life. At the same time, we've come to understand that the range of environments where life can survive is bigger than once we thought. Without leaving our own planet, we have found life under conditions where survival seems impossible—but where living beings have evolved and thrive, sometimes in forms so strange they might well have come from another world.

The most fascinating of these exotic sites were found in 1977 under a mile or more of water. These are the undersea geothermal vents; that is, cracks in the ocean floor that extend down to the magma, the hot molten rock inside the Earth. Seawater entering the cracks is heated to over 700 degrees Fahrenheit but doesn't boil because of the immense pressure. Instead, the hot water is ejected in plumes at the ocean bottom, so full of sulfur compounds and other minerals that it sometimes turns black, which is why the vents are called "black smokers." These hot spots, where the ocean bears down with vast weight, where water is heated beyond normal boiling, where toxic chemicals abound, and where the sun never shines, could serve as an updated vision of Hell for humans, but not for the previously unknown life-forms that have arisen and flourish around the vents.

Some of this vent life is microscopic, bacteria that survive temperatures of 300 degrees Fahrenheit, whereas boiling at 212 degrees Fahrenheit (100 Celsius) kills most microorganisms. Then there are the vent worms, up to six feet long, which lack a mouth and a digestive system. Instead, they're loaded with bacteria that carry out the trick of converting hydrogen sulfide from the plumes into energy for the worms. Plants on the Earth's surface perform photosynthesis to change sunlight into cellular energy; at the sunless vents, these bacteria perform chemosynthesis instead. Very recently, researchers have also found bacteria that seem to perform photosynthesis using faint light generated near the vents. No one is sure how the light is made, but if the discovery is confirmed, it would be the first life-form we have discovered that lives off a light source other than the sun. Such findings about vent creatures offer clues about how life arose on our planet and show possible new pathways for life on other planets.

Other earthly life-forms also expand the boundaries of possibility for life to do well under all kinds of extreme conditions—not only great heat, but extreme cold, high acidity or saltiness, and enough radiation to kill a human 500 times over. Some microorganisms live two miles underground; others have survived years in the vacuum of space aboard a NASA satellite. This tremendous diversity enlarges our ideas about where life might be found. If there's no life on top of the Martian deserts, maybe it exists beneath them. Maybe Europa, with liquid water and a hot interior, supports geothermal vents and associated life, or its beginnings. Maybe those extrasolar hot Jupiters differ enough from the cold Jupiter we know to allow life a toehold, if a very different one from anything we now expect.

Adding to these possibilities, the gold standards for the existence of life, abundant carbon compounds and liquid water, can perhaps be relaxed. Like carbon, atoms of silicon—the element used in computer chips—have four chemical bonds and can form complex molecules. This possibility is recognized in *Alien*, where Science Officer Ash points out that silicon forms a kind of protective armor for the monster's early stage. However, silicon-based life would have problems compared to carbon-based forms. The combination of carbon with oxygen, CO_2, the result of respiration by living things, is a gas that can easily be expelled by an organism—it's what we exhale as we breathe. But the analogous form for silicon, SiO_2, is a solid, which can't be so readily removed. Also, silicon compounds are less prevalent in the universe than the carbon varieties.

Water, too, might not be absolutely essential for life, because it isn't the only liquid to support complex chemistry. Liquid ammonia, a compound of nitrogen and hydrogen (NH_3), also dissolves materials to allow a rich range of interactions. Further, it remains liquid at lower temperatures (between –108 degrees Fahrenheit and –27 degrees Fahrenheit) and higher pressures than water, so an ammonia-based biochemistry could support life under more extreme conditions than a water-based one. Another possibility is water laced with ammonia, which would lower the freezing point of the water and may exist on Titan.

While we're looking for alien life, we must ask if alien life is looking for us. Some people believe aliens have already found us, visiting the Earth in UFOs like the alien space craft examined in *Independence Day* and abducting humans for further study. But we don't need dubious UFO scenarios to imagine a way to connect with distant aliens.

The reasoning goes like this: without even meaning to, we humans have been sending out signals to the rest of the universe for a century, in the form of ordinary radio and television waves that head out to space. (The movie *Contact* [1997] opens with a scene showing radio and television programs from the past as they travel out from Earth.) Any advanced alien civilization would be expected to have the same technology, because electromagnetism is a basic phenomenon of the universe. Thus there may be other races sending out signals—either deliberately or as a cultural byproduct—that carry some kind of intelligence, whether that be mathematical formulas or otherworldly versions of *The Simpsons* or *As the World Turns*. If we listen to what's coming at us from the sky, we might detect aliens without leaving our planet.

It isn't easy, though, to search for intelligent messages. The universe is full of natural radio signals generated inside stars and by other processes. To further confuse things, some of these repeat at a steady rate, suggesting an artificial rather than a natural source, so one major problem is separating meaningful signals from cosmic noise. There are other uncertainties too. Given the vast number of stars, in what direction do you listen, and at what radio frequency?

Nevertheless, in an act of faith, the astronomer Frank Drake started listening to the universe in 1960, before we even knew there were extrasolar planets out there. He used a large radio telescope, which is really an ultrasensitive radio antenna. He also developed what's called the Drake equation, which roughly estimates the number of alien civilizations in our galaxy that might be broadcasting signals. Some of the quantities that go into this estimate, such as the number of years that an advanced civilization could be expected to survive, are highly speculative because we simply don't have much data. Optimistic interpretations of the equation predict dozens to hundreds of candidate civilizations in our galaxy, but pessimistic evaluations say there wouldn't even be one.

Still, the SETI project (Search for Extraterrestrial Intelligence) that Drake started has listened to radio waves from space for nearly fifty years. So far, no signals have been detected that can be identified as coming from an intelligent source. Even so, SETI represents something unique. Although we can successfully explore the solar system, sending spacecraft to planets orbiting other stars is an entirely different proposition because of the enormous distances. At the speed of a NASA spacecraft, it would take an interstellar ship over 100,000 years to reach the nearest star, Alpha Centauri, and over 500,000 years to reach Gliese 581d, the planet twenty light-years distant that lies in the habitable zone. Even at the speed of light, which anyway is theoretically impossible for a spaceship, it would require years and centuries to reach distant stars. Though SETI has only a slim chance of intercepting alien radio signals, it is for now the only way we can seek extrasolar civilizations.

Whether we come to the aliens or they come to us, what would they look like, and how would they act? While scientists would go wild with excitement over any alien life-forms whatever, one possibility is that for the casual onlooker, alien life would be nothing to write home about. It could be invisible to the naked eye, like Martian bacteria. It could be primitive and immobile, like lichen, and equally unthrilling to observe. But if we do hit the jackpot and connect with fully developed, thinking, moving alien creatures, are the movies good guides to what we might expect?

Some alien behaviors we've seen in the movies might not surprise us if they actually showed up, because they have analogues in Earthly species. For instance, the scene in *Alien* where the creature bursts out of Kane's stomach is a real event on Earth, though, thankfully, not with people. Certain types of wasps plant their eggs inside particular insect larva or caterpillars. When the eggs hatch, the result is a miniversion of the horror scene in the film.

Similarly, parasites that live on or in a host of a different species, as shown in *The Puppet Masters*, are common on Earth for various species, including humans. Body and head lice have been with humanity throughout our history. More serious afflictions come from tapeworms and other worms that live within the digestive tract. These cause illness and absorb nutrients and are thought to affect over a billion people. Parasitic protozoa, single-celled organisms that can take up residence in various parts of the human body, also make people ill. But in contrast to the slugs in *Puppet Masters*, which waltz onto our planet and seize control of humans—alien to them as they are to us—instantaneously and with scarcely a hiccup, a parasite must coevolve over time with a host before it can benefit from that host. And we certainly don't know of any earthly parasites that exert mind control.

Another unusual alien characteristic, plant mobility, as represented by the creature in the original *The Thing from Another World*, isn't unknown on Earth either. Some plants display surprisingly complex and rapid motion when they disperse seeds and trap insects. But plant movement uses different mechanisms than the muscular tissue that animals employ. A recent analysis by Jan Skotheim and L. Mahadevan, of Cambridge and Harvard Universities, shows that these mechanisms are not of a scale and speed that would enable the Thing to stride down corridors, break through doors, and battle sled dogs—not to mention the exact details of how a plant that begins in a rooted manner, as the movie shows, becomes free to roam.

The idea of emotionless plants as meaningful and even superior alternatives to humans, expressed by Dr. Carrington in *The Thing*, also needs updating. We tend to believe in a strict division between rational and emotional thought, with emotions often interfering with rationality. Carrington articulates an extreme version of this, wanting to completely eliminate emotion to exercise pure, untainted reason. But now there's considerable evidence of important connections between the part of the human brain associated with rational thought, the cortex, and the part associated with emotion, the limbic system. Physiologically speaking, you can't really separate the two. Some researchers, such as the neurologist Antonio Damasio, think that certain sophisticated cognitive functions could not work without emotions; for instance, it's been shown that lack of emotion can make a person literally incapable of choosing from a menu of alternatives in a valid and "rational" way. So it might be that the emotionless plants in *The Thing from Another World* and the unfeeling human replicas in *Invasion of the Body Snatchers*, along with Dr. Carrington himself, aren't really all that smart, and that advice from ultralogical Mr. Spock in *Star Trek* isn't always the way to go either—as Captain Kirk seems to intuitively understand.

If the movie versions of existing Earthly behaviors are sometimes amplified and exaggerated, other alien abilities are completely made up. Many people have personal anecdotes about telepathic moments, but there is as yet no convincing scientific evidence that direct mind-to-mind communication exists, although such a connection is implied in the 1956 *Body Snatchers*, and President Whitmore in *Independence Day* has brief telepathic contact with one of the aliens.

Paradoxically, to balance those cases where alien capabilities are pure speculation, sometimes movie aliens follow earthly life too slavishly. Many aliens in film, even those meant to be ferocious, grotesque, or disgusting, are remarkably humanoid in general bodily outline and facial details. Usually the alien body stands upright. It has a torso and a head with two eyes (the Martians in the original *The War of the Worlds* are an exception). It has two arms and two legs as well, arranged in bilateral symmetry, that is, the body's left and right halves are mirror images. The beings in *Independence Day*, for instance, have this contour, although with the evil devilfish shape of the head, and with tentacles added to the mix, you still wouldn't want to pal up with one at your favorite bar.

Some films assume that the similarities between us and alien life go deeper than bodily shape. One issue is interspecies infection, as happens on Earth, for example, with rabies and the HIV virus. In all versions of *The War of the Worlds*, the Martians go down because they're susceptible to earthly bacteria, which implies anatomical similarities between the two species. In *The Puppet Masters*, the slugs are defeated by infecting them with an Earthly virus, but this would work only if the cells making up the alien bodies function similarly to human cells, because viruses reproduce by taking over the internal cellular machinery.

No one knows for sure if infection in either direction—human to alien, or alien to human—can really happen. But NASA took the possibility seriously enough to put the astronauts who landed on the Moon into isolation for twenty-one days when they returned, and the agency now runs a "planetary protection" program to sterilize spacecraft that land on other planets. Many scientists think that these protective efforts should be made far more stringent, especially if we retrieve what might be life-bearing samples from Mars or elsewhere and bring them back to Earth.

Another assumption sometimes made is that we can all breathe each other's air. In *Starship Troopers*, human soldiers on the Bug home planet Klendathu fight the enemy just fine without spacesuits, and a captured Bug remains murderously active on Earth. On the other hand, in the 1953 *War of the Worlds*, the scientists comment that the Martians might have problems in our atmosphere. The odds that different planetary atmospheres would be as similar as portrayed in *Starship Troopers* are probably slim, but sometimes similarity is essential to the plot. The Bugs come into conflict with humanity because the aliens are seeking planets where humans could survive as well. And from the cinematic viewpoint, putting actors into spacesuits is a problem because that makes it hard to show reactions and facial expressions.

Then there are other movie aliens, like the Bug warriors and the repulsive slugs in *The Puppet Masters*, that look and act anything but human. While there's no telling if aliens would be quite that repellent, according to how evolution works, it's nearly impossible that they would look just like us or any Earthly species. There's every reason to think that organisms would develop on other planets according to the same evolutionary laws that work on ours, and how evolution plays out depends on the physical and biological environment.

The environment we know best, of course, is our own, and the chances of life arising on Earth-like planets are especially strong. According to the late Harvard paleontologist and writer Stephen Jay Gould, "The origin [of] life seems reasonably predictable on planets of earthlike composition . . . I don't know how else to interpret the cardinal fact that life did originate on earth almost as soon as environmental conditions permitted such an event." But even on a more or less Earth-like planet, small changes would lead to a very different evolutionary story. In *The Science of Aliens*, Clifford Pickover summarizes the dependence of life on the physical environment: "evolution is so sensitive to small changes that if we were to rewind and play back the 'tape' of evolution, and raise the Earth's initial overall temperature by just a degree, humankind would not exist . . . if humans were wiped out today, humans would not arise. This means that on another world, the same genetic systems and genes will not arise."

As an example of this sensitivity, consider the sunlight that bathes our planet. Our eyes have adapted to work efficiently with the wavelengths that make up that light. These are determined by the sun's surface temperature, about 9,900 degrees Fahrenheit or 5,500 Celsius. Inhabitants of a planet circling a sun a few hundred degrees hotter or colder would experience respectively more blue or more red light in the mix, with a corresponding change in adaptation. And if that sun's surface temperature were drastically higher or lower, any planetary inhabitants would have evolved to see by ultraviolet or infrared light, not visible light, as determined by the wavelengths the sun emitted.

If the planet differed from the Earth in mass and size, that would produce a different gravitational force, another important factor. On Mars, with its weaker gravity, living creatures could manage with a spindlier construction than on Earth, as expressed in the *War of the Worlds* stories. Other planets have stronger gravity. If you weigh 100 pounds on Earth, you would weigh 250 pounds at the outer edge of Jupiter's gaseous globe. If Gliese 876d, the distant extrasolar planet seven times as massive as the Earth, turns out to have the same rocky composition as the Earth, its gravity would pull twice as hard as ours. Beings that developed in such a high-gravity environment might have a low-slung bodily structure with thick limbs so they could stand and move against that relentless pull.

Still, it's probable that some universal features would show up across different evolutionary tracks, an idea called convergent evolution. Certain problems posed by the environment can be solved only in certain ways; for instance, different kinds of creatures that need to move swiftly through water to catch their prey, say, a seal and a shark—one a mammal, one a fish—have developed streamlined shapes to slip through the medium with minimum effort.

But although beings that evolved elsewhere might share some convergent aspects with Earth species, there's no reason to think they would closely resemble anything we know here, and they might seem very strange indeed. As Pickover puts it: "The enormous diversity of life today represents only a small fraction of what is possible." Even on our own planet, animals like the vent worms show that creatures from different ecological niches can be utterly alien. There are countless other examples, such as the octopus. Also, creatures from earlier evolutionary eras differ greatly from what flourishes today, the best known example being the enormous variety of dinosaurs that lived more than 65 million years ago, before mammals became important.

To show how alien-seeming creatures can arise as evolution marches on, consider a fossil called *Hallucigenia sparsa*, found in the Burgess Shale. The Shale, once under the sea, now sits high up in the Canadian Rockies. It's a rich fossil bed from the Cambrian Period, a time over 500 million years ago that was an amazing era in the history of life on Earth. An incredible variety of creatures, more complex than their predecessors, appeared in the explosively short time (by geological standards) of 20 million years. The Shale preserves many of these species; some are forebears of plants and animals that still exist, others seem unrelated to anything now alive. Because of the nature of the fine mud that covered these life forms, the fossils give detailed information about soft body parts as well as the hard parts that are more usually found.

Among the Burgess fossils, *Hallucigenia sparsa*, whose name reflects its weirdness (*hallucigenia* means "source of dreams"), stands as a symbol for life's diversity. The creature has a banana-shaped body, with no sign of sensory organs or other indication as to which end is the head. Along one side of the body grow two rows of spiky needle-like appendages, seven in each row. In the first fossils recovered, the opposite side of the body supported a single row of seven flexible tentacles, with some sort of structure at the end of each. That led to an interpretation that had the creature walking on its twin rows of spikes. The seven tentacles sprouting from its back were thought to terminate in mouths for feeding.

But examination of more fossils showed that the creature really carries two rows of tentacles. Now, a newer interpretation turns *Hallucigenia* upside down; it seems that it walked on its seven pairs of tentacles, each tipped with a claw. In this configuration, the spikes lined its back, perhaps to protect it against enemies. Either way, right side up or upside down, if you saw this thing approaching you, at say the size of a dog (the fossils are an inch or less long), you would have no doubt that it came from somewhere far, far distant from Earth. In fact, a low-slung body resting on fourteen limbs could work well for a heavy gravity planet.

This real-life example of a very strange creature shows that the portrayal of aliens in film has barely scratched the surface of the weirdness that might lie out there. As movie special effects become more flexible and more persuasive, and as we further understand how alien life might develop, we'll move away from depictions based on people in makeup and monster suits, and toward truly exotic movie aliens that are also scientifically valid, or at least possible.

In terms of what science now tells us about alien life, the significant point is this: whatever we once thought about the probabilities of finding life elsewhere, those chances have increased enormously and dramatically in the last few years. What Scotty the reporter says at the end of *The Thing from Another World* is better advice and rings truer than ever we knew in 1951: "Keep watching the skies."

As the next chapter shows, that means watching not only for aliens but for cosmic objects hurtling in from space that might hurt us and our world.

Chapter 3

Devastating Collisions

Dave Randall: Never mind. Good air or bad, it's the only place we can go.

—*When Worlds Collide* (1951)

Narrator: This is the Earth, at a time when the dinosaurs ruled. . . . A piece of rock just six miles wide changed all that. . . . It happened before. It will happen again.

—*Armageddon* (1998)

Mark: How do we set the nukes inside the comet and get out before they blow?
Orin Monash: We don't.
Andrea: Look on the bright side. We'll all have high schools named after us.

—*Deep Impact* (1998)

If you were a regular viewer of science fiction films anytime from the 1950s to the 1990s, you might have felt vulnerable just standing on the Earth's surface. According to films like *Meteor* and *Deep Impact*, you could expect some immense piece of space debris, the size of Mt. Everest or bigger, to come smashing into our planet every so often. It would raise havoc, causing earthquakes and tsunamis, wiping out cities, and threatening to destroy all earthly life. Also according to these films, although you wouldn't be safe anywhere, big cities were the worst place to be. Against all odds, the missiles from space seemed to select urban areas for particular devastation. New York was a favorite, but there was little safety to be found in Paris or Buenos Aires either.

These movie scenarios rely on the fact that our solar system contains millions of pieces of space debris, some of which have crashed into the Earth in the past. Some are crashing into it right now, up to one hundred tons a day (or at least, this much is reaching the atmosphere, where most pieces burn up), and more will do so in the future. Even a smallish incoming celestial object like a rocky chunk of asteroid, hitting the Earth at thousands of miles per hour, carries the destructive power of an atomic or hydrogen bomb; a bigger chunk can equal the effects of thousands of bombs.

You can see the scars of these cosmic accidents with your own eyes: for instance, by visiting Barringer Crater in Arizona, a hole nearly a mile across blasted out by a much smaller incoming object; or by gazing through a telescope at the Moon's craters, produced by past collisions. The Moon itself is thought to be the result of the biggest collision our Earth has ever known. Billions of years ago, an object the size of the planet Mars struck our planet a glancing blow. That terrific smack tilted the Earth on its axis, which is why we have the four seasons of the year. It also destroyed the impacting body and stripped off the Earth's outer layer, providing a stockpile of raw material orbiting the Earth. Eventually, gravitational attraction among these pieces brought them together to form the Moon.

The Earth has suffered other major collisions, one of which is thought to have wiped out the dinosaurs 65 million years ago. These creatures had dominated the Earth for 150 million years, until an asteroid slammed into the edge of the Yucatan Peninsula in Mexico, starting a chain of events that led to their demise and the extinction of other life-forms. Only a relatively small number of species survived, some of them evolving into modern mammals and, eventually, humans.

The possibility of another, similar impact has not gone away. As recently as 1994 and 2002, our planet had near misses with asteroids. Also in 1994, astronomers and the general public watched, mesmerized, as pieces of a comet collided with the planet Jupiter. But if movies are not wrong in showing cosmic collisions, are the effects as dreadful as the films suggest? In the movies, the devastation is never anything short of awesome. It has been known to go to the limit, the death of all living things on Earth, and in the classic catastrophe film *When Worlds Collide* (1951), the destruction of the Earth itself.

When Worlds Collide is based on a novel of the same name by Philip Wylie and Edwin Balmer. The film's producer was George Pal, who had turned out

Destination Moon and would go on to *The War of the Worlds*. In the story, astronomer Dr. Cole Hendron (Larry Keating), of Cosmos Observatory, warns the world that two rogue cosmic bodies named Bellus and Zyra, traveling together, are headed for Earth. Zyra is a planet similar to ours, perhaps even with vegetation. Bellus is a star twelve times the size of the Earth (in present astrophysical theory, that would mean Bellus is too small to be an active, energy-producing star, although it could be a dead one). Zyra will sweep past the Earth in less than a year, when it will disrupt land and sea, causing earthquakes, volcanoes, and tsunamis. And if this isn't bad enough, nineteen days later, Bellus will provide the coup de grace, smashing directly into the Earth to finish it off.

Hendron's warning isn't taken seriously by the United Nations, but millionaire industrialist Sydney Stanton (John Hoyt) bankrolls Hendron in a daring scheme to save remnants of humanity including Stanton himself. The plan is to build a space ark that will carry a handful of people to Zyra before Bellus hits, along with supplies to begin life on the new planet. The spaceship, a sleek model with fins, will be launched along a mile-long track.

With the help of Hendron's daughter, Joyce (Barbara Rush), aviator-turned-spaceship-pilot Dave Randall (Richard Derr), and an army of workers, a frantic construction effort gets the ship built just in time. Despite bad moments during the destruction induced by Zyra, and riots by those not chosen to board the spaceship, the craft takes off successfully. (The devastation caused by Zyra, such as New York's streets filling with water and ships floating among skyscrapers, is shown via special effects that were extraordinary for those days and earned the film an Academy Award.) In the final scenes, forty survivors land safely on Zyra, which luckily turns out to have breathable air along with plants and water. The film closes with a crescendo of uplifting music and the on-screen legend "The first day of the new world had begun . . ."

Nearly thirty years later, the movie *Meteor* (1979) offered a seemingly better solution than a space ark. This time the troublesome object comes from within our own solar system. It's been knocked loose from the asteroid "Orpheus" by a collision with a comet and by sheer bad luck makes a beeline for Earth (after a smaller piece of the asteroid wipes out a NASA spacecraft and its crew). Although not the size of a planet as in *When Worlds Collide*, this massive piece of rock is quite a chunk; it's some five miles wide.

When Worlds Collide (1951). Ocean liners float amidst Manhattan's flooded towers as the gravitational pull of the passing planet Zyra produces global catastrophe—and that's even before the accompanying star Bellus crashes into the Earth. Special effects won this film an Oscar, but the film forgets that the Earth's gravity would produce similar bad effects on Zyra, making it less than welcoming for the few surviving humans.

Source: Paramount/The Kobal Collection.

This is big enough to wipe out humanity, as spelled out by former NASA scientist Dr. Paul Bradley (Sean Connery). The rock, roaring in at 30,000 miles per hour, will hit the Earth with an explosive energy like that from two million megatons of TNT. This, he says, will produce a concussion ten billion times as powerful as the biggest earthquake ever, lofting billions of tons of earth and dust into the air. That blanket will block the sun's rays and lower the Earth's temperature for decades, killing plants and eventually human and animal life.

The only way to stop this catastrophe is to attack the rock with nuclear weapons. The United States had put nuclear-tipped missiles into Earth orbit for just this possibility, but then turned the weapons toward its Cold War enemy, the Soviet Union, which is why Bradley left NASA. But although the missiles can be redirected toward the incoming threat, they don't have enough power to destroy it. After much political maneuvering, the Soviets agree to add their own orbiting weapons aimed at the United States to the anti-asteroid arsenal, and the two countries send a joint flight of missiles toward the rock. Before these can strike home, "splinters" from the rock arrive and cause damage around the world: Hong Kong is flooded by a hundred-foot tidal wave; the Swiss Alps suffer immense avalanches when a splinter takes out the top of a mountain; and Manhattan is left in shambles. Amid this devastation, though, the American and Soviet missiles fly true and destroy the rock, thus avoiding the biggest devastation of all.

After *Meteor*, giant collisions on film held off through the 1980s, although a remote cousin of the idea appeared in *Night of the Comet* (1984). The story combines science fiction and horror elements as two California Valley girls find themselves among the few survivors after the Earth passes through the tail of Halley's Comet, and they spend the rest of the film fighting off flesh-eating zombies. But massive collisions returned with a vengeance in the 1990s. Inspired by new special effects that enhanced on-screen depictions of giant waves and exploding asteroids, two films about collisions that could destroy all Earthly life appeared in 1998, *Armageddon* and *Deep Impact*; in 1997, *Starship Troopers* envisioned a lesser but still awful scale of destruction from space, the deliberate wiping out of a city.

Deep Impact was originally meant to be a remake of *When Worlds Collide*. It evolved differently but kept the idea of the Earth facing two space objects, although only one, a comet, shows up initially. As described in the solemn tones of the U.S. president (Morgan Freeman), this body is the size of Manhattan from the Battery to the northern edge of Central Park—that is, seven miles across—and weighs 500 billion tons. Following the same scenario as in *Meteor*, the United States and Russia together try to destroy the rock. Crusty, experienced, fly-by-the-seat-of-the-pants astronaut Spurgeon "Fish" Tanner (Robert Duvall) and his crew land on the nucleus, or solid part of the comet, in the spacecraft *Messiah*, implant nuclear devices 300 feet down, and detonate them. But instead of smashing the comet to bits, the explosions only crack it into two big pieces. Unfortunately, both keep right on coming at the Earth.

The smaller chunk, still a major lump 1.5 miles across, arrives first, landing in the Atlantic Ocean off North Carolina. Making one of the biggest

Deep Impact (1998). A panicky crowd, trapped in a monumental traffic jam, watches as a piece of comet more than a mile across heads for the Atlantic Ocean off Cape Hatteras. They're trying to escape the massive destruction that will result as the huge walls of water raised by the ocean strike penetrate far inland. Although an ocean impact might seem less violent than a land impact, it could produce great devastation.

Source: Dreamworks/Paramount/The Kobal Collection/ILM (Industrial Light & Magic).

splashes ever, this raises a tsunami that becomes thousands of feet high as it approaches land, which means it would tower over the world's tallest skyscrapers. *Deep Impact* shows a marvelous scene where this gargantuan wall of water engulfs Manhattan and causes widespread devastation.

But the real problem is still in the mail; that's the larger body, at six miles across plenty big enough to qualify as an "extinction-level event." Amid worldwide rioting as society begins to break down, a second effort to destroy it, with ICBMs, also fails. The world seems doomed, but then humanity's luck changes. Returning to Earth in the *Messiah*, Fish Tanner sees that the sun's heat has vaporized the comet's ice to make a deep hole. Setting off the remaining nuclear weapons in the hole would blast the comet to pieces. But there isn't enough fuel to fly the *Messiah* into the hole and out again, so with the consent of his crew, Fish flies in on a one-way suicide mission. The plan works and the weapons explode the comet into pieces without damaging the

Earth. Although tsunamis from the first object have killed millions and left destruction in the United States, Europe, and Africa, the film ends with a sense that humanity will rebuild.

True to its title, *Armageddon* outdoes both *Meteor* and *Deep Impact* with an even bigger rock. The first bad news comes as swarms of space objects destroy a NASA shuttle and cause significant damage, from the east coast of the United States to Europe. A bigger piece hits Shanghai, killing 50,000 people. These events are the forerunners of mega-bad news: a truly enormous asteroid is due to hit the Earth. Forget the city-sized rock in *Deep Impact*; this one is as big as Texas, a "global killer" that is more than enough to finish off every living thing on our planet, even bacteria.

As in *Meteor* and *Deep Impact*, the hope for salvation is to disrupt the asteroid with nuclear weapons. A NASA scientist calculates that a bomb detonated deep within the asteroid will break it in two pieces, which will fly off on diverging courses that miss the Earth (unlike what happened in *Deep Impact*).

Armageddon (1998). Before a really big asteroid smashes into the Earth and wipes out all life, a smaller space rock scores a direct hit on Manhattan's Chrysler Building. The odds against an object from space hitting any particular spot on our planet are incredibly high; however, such objects really have collided with the Earth, sometimes catastrophically.

Source: Touchstone/The Kobal Collection.

There isn't time, however, to train astronauts in rock-drilling techniques, so the plan is to rapidly train a team of oil rig drillers (Bruce Willis plays their boss, Harry Stamper) to operate in space.

The drillers are sent up to the asteroid in two NASA space shuttles, but things don't go smoothly: a refueling stop at a Russian space station almost turns disastrous, a crash landing on the asteroid kills most of one shuttle crew, and drilling conditions are extraordinarily difficult. Still, the team manages to tunnel down 800 feet and plant the bomb. Unfortunately, its remote detonator has been damaged, and it can be set off only by hand. With seconds to go, Stamper pulls the trigger, and the asteroid splits into pieces that fly off harmlessly. Humanity is saved at the cost of Stamper's life.

A different spin on the planet-busting, all-life-destroying, random type of collision comes from *Star Ship Troopers*. As part of their war against humanity, the film's hostile aliens, the Arachnids, or Bugs (who appear in chapter 2), wipe out the city of Buenos Aires with what might seem a primitive weapon. Instead of nuclear bombs or giant lasers, they simply throw a rock at the city, lobbed in from their home planet Klendathu.

The devastation is total: Buenos Aires is in flames, its buildings are rubble, and the casualty count is 21 million dead and wounded. (Present-day greater Buenos Aires has a population of 13 million, so the city has grown enormously by the time the story takes place.) The narrator adds ominously that Klendathu lies near an asteroid belt that contains many more rocks ready to be hurled at Earth. In fact, the rest of the film focuses not on space bombardment but on how humanity's Mobile Infantry takes the fight to Klendathu and defeats the Bugs.

Whether in deliberate military use or in a natural disaster, it's hard to see how mere chunks of rock, metal, and ice can cause such damage and death. Although a rock weighing 500 billion tons is a serious piece of matter, it's only a tiny fraction of the Earth's mass. Could its impact really wipe out everything living, and could a smaller chunk really take out a city? Could a near miss by another planet cause major destruction, as in *When Worlds Collide*, and could a smaller object raise a tsunami half a mile high, as in *Deep Impact?* To put it bluntly, was all the movie panic necessary, or was it overblown, just another way to sell tickets without worrying about scientific accuracy?

Incredible as it may seem, there's a lot of truth in collision films, from *When Worlds Collide* onward. Massive, fast-moving objects from space can dig enormous craters, raise giant waves, and threaten whole populations. If they're big enough, they can threaten life on our entire planet. But do such potentially disastrous missiles even exist anywhere near the Earth? The answer is yes, and in large quantities, in the form of asteroids, comets, and meteoroids—all different forms of space objects flying through our solar system.

Like the Bug planet Klendathu, our solar system possesses an asteroid belt—a collection of chunks of rock and metal that orbit the sun mostly between the orbits of Mars, the next planet out from the Earth, and Jupiter. The first asteroid ever sighted is the biggest one known: Ceres, discovered by the Sicilian astronomer Giuseppe Piazzi in 1801 and named after Sicily's patron goddess. With a diameter of nearly 600 miles, it's a quarter the size of our moon. (Like Pluto, Ceres has been redefined; in 2006, it became a dwarf planet, the smallest in the solar system, rather than an asteroid.) Since then, 165,000 asteroids have been positively or tentatively identified, but only about 200 of these are bigger than 120 miles across. There are, however, estimated to be millions of relatively small asteroids less than 0.6 mile across.

The asteroids were once thought to be remnants of an exploded planet that had occupied the gap between Mars and Jupiter, an idea that lent itself to stories about an alien race that wiped out its own planet. Now there's a more sober scientific assessment; asteroids are not the shattered remains of a planet, but leftover building blocks from the early days of our solar system. In that time, nearly five billion years ago, the gas and dust that had collected in our corner of space began to clump together under the force of gravity. As gravity worked its magic, tiny pieces of matter came together into bigger chunks of rock and metal called planetesimals.

Most of these eventually combined further into whole planets, but not all. The planetesimals near Jupiter were disturbed by the planet's strong gravity and remained spread out to form the asteroid belt. The bodies in this belt, even big ones like Ceres, don't threaten us; they move around the sun in nearly circular orbits and never come near the Earth. But some asteroids have elliptical orbits that swing them alternately closer to the sun and further out, so that they cross the Earth's orbit.

About 1,700 of these Earth-crossing bodies are known, called collectively the Apollo asteroids. One could conceivably cross our orbit just as the Earth reaches the same point, with an inevitable collision. We've had recent near misses with some of the Apollos. In January 2002, an asteroid missed us by 230,000 miles—the distance to the Moon, our closest neighbor in space. In December 1994, another asteroid named 1994 XM1 came within 60,000 miles of the earth, but the closest call of all came in 2004, when asteroid 2004 FH approached within 27,000 miles. That's uncomfortably near, only slightly further than the 22,300 miles at which many artificial satellites orbit the Earth. Each of these near-miss asteroids measured about 100 feet across, big enough to do damage if it landed in a vulnerable place.

One particular asteroid, much larger at 0.7 mile across, truly has a chance of hitting the Earth in the near future. Asteroid 1950 DA was discovered in 1950, lost, and rediscovered in 2000. The original sightings were supplemented with radar tracking by J. D. Giorgini of NASA's Jet Propulsion Laboratory and his colleagues, leading to an unusually accurate orbital determination. This predicts that 1950 DA will have a 1/300 or 0.33 percent probability of hitting the Earth on March 16, 2880. One in 300 is a very slim chance, but much higher than that for any other known bodies. Further radar observations might refine the numbers, but according to Giorgini and his colleagues, as 1950 DA gets nearer, it might require action right out of a science fiction movie, direct inspection by a spacecraft, to decide if it really will hit the Earth.

Comets are also pieces of the original solar system, formed near its outer fringes, where it was cold enough to freeze water and other compounds. Orbiting pieces of rock and metal combined with ice to form comets, billions of which now occupy two reservoirs in space: the doughnut-shaped Kuiper belt and the spherical Oort cloud (each named after an astronomer). Both lie beyond Pluto, but many comets have orbits that are long skinny ovals. These bodies alternately swing far out into space, and penetrate the inner solar system. That explains why Halley's comet, like others, is out of sight for long periods, but swims into view on a regular basis—every seventy-six years.

Although the most spectacular feature of comets is their tails (each comet has two, one of gas and one of dust, released from the comet's ice by the warmth of the sun), it's the solid core of a comet, its "dirty iceberg" of a nucleus, that can be dangerous. For instance, Halley's Comet's nucleus is nine miles long by five miles wide. A comet nucleus is less massive than an asteroid of the same size because ice is less dense than rock or metal, but it can still cause serious damage. This was dramatically apparent when some twenty pieces of comet Shoemaker-Levy 9, two miles across or less, hit Jupiter in 1994. The impacts produced giant fireballs hotter than the surface of the sun and massive atmospheric disturbances.

Besides asteroids and comets, there are meteoroids—smaller bodies such as pieces of broken asteroids cruising through the solar system. Many are very small, less than 0.04 inch, or 1 mm, across, like grains of sand. Even small ones can be spectacular if they enter the Earth's atmosphere (at which point, like any space object entering our atmosphere, they're properly called meteors). The friction of a meteor's passage generates enough heat to vaporize its outer skin, leaving a streak of hot gas, which we call a shooting star. A small meteor will vaporize completely before hitting the Earth. With bigger ones, and with comets or asteroids, enough material is left to reach the ground, and any surviving piece is called a meteorite.

While there's plenty of ammunition that might hit our planet, are these chunks as destructive as the movies suggest? The answer is "yes," because like any moving body they carry kinetic energy, and in huge amounts. A body in motion has energy because when it rams into something, it exerts a force that makes things happen—whether the moving body is a falling hailstone that dents a car, a hurtling locomotive that crumples an SUV at a grade crossing, or a mountain-size rock from space that digs a crater.

The amount of kinetic energy depends on the mass or amount of matter in the body, and on its speed. A 500-billion-ton comet like that in *Deep Impact* would have enormous kinetic energy even creeping along at a walking pace. Kinetic energy is proportional to mass; double the mass of an object and you double its energy. But kinetic energy *really* goes up if the object is moving fast. If you double the speed, the object has four times the energy; triple it, nine times the energy, and so on.

Like all bodies in space, objects that might crash into the Earth are moving at high speeds in their own orbits. Also, the Earth's gravity accelerates an object as it approaches, although some speed is lost as it plows through our atmosphere. Typical speeds as a space object nears the Earth are far above anything in our daily experience, or even those reached by NASA spacecraft: incoming space objects are moving at 38,000 to 110,000 miles per hour (mph) or more, with the lower values applying to asteroids and the higher ones to comets.

High speeds combined with huge masses translate into enormous destructive power, although calculating the catastrophic effects isn't an exact science.

Scientists haven't seen enough big real-life collisions to be absolutely sure about all the parameters, but they understand the process well enough to draw fairly accurate conclusions. For one thing, it's startling to realize how small an object it takes to blast a big hole, typically twelve to twenty times the diameter of the object. The Barringer Crater in Arizona, 0.8 mile across and about 600 feet deep, was gouged out 50,000 years ago by an object made mostly of iron and only around 150 to 180 feet across (half a football field), barreling into the ground at 25,000 mph or more.

Another way to measure destructive power is in comparison to nuclear weapons. When the United States dropped an atomic bomb on Hiroshima in 1945, the weapon was rated at fifteen kilotons, that is, it carried the explosive power of 15,000 tons of TNT. The bomb dropped on Nagasaki three days later was a twenty kiloton weapon. The two weapons illustrated Einstein's famous equation $E=mc^2$, which says that small amounts of matter can be converted into enormous outpourings of energy. Each bomb generated the staggering destructive power of thousands of tons of high explosive by converting only a few pounds of uranium, in one case, and plutonium, in the other, into energy.

These A-bombs caused 100,000 or more deaths outright (from a combined population of 450,000), as well as the physical devastation of two cities; yet the next level of nuclear weapon, the hydrogen bomb, has unimaginably greater power. The first H-bomb, built by the United States and tested at Eniwetok Atoll in 1951, was rated at several *million* tons, or megatons, of TNT. Eventually the United States and the Soviet Union were stockpiling H-bombs of ten to fifty megatons; each was the equivalent of 500 to 2,500 Hiroshima atomic bombs, enough to destroy big cities and whole regions.

Really big asteroids or comets, like the five-mile asteroid in *Meteor*, up the ante. They can deliver millions of megatons of destructive power, just as Paul Bradley announced in the film: like thousands of hydrogen bombs exploding at one spot to cause regional or global devastation. In real life, the biggest fragment of comet Shoemaker-Levy 9 delivered some 600 million megatons when it crashed into Jupiter in 1994. Even the infinitely smaller 150-foot object that made the Barringer Crater carried the energy of a ten- or twenty-megaton hydrogen bomb, enough to inflict nuclear-bomb-like death and damage on a major city. It would be the worst kind of luck for a space object to land on a city, since incoming missiles are equally likely to hit anywhere on

Earth. Still, to get an idea of the magnitude of an asteroid strike, we can imagine it happening in an urban setting.

If a 150-foot iron object were to smash into Manhattan at Forty-second Street and Fifth Avenue, it would create a mile-wide crater, extending north-south from Thirty-second St. to Fifty-second St. and east-west across much of Manhattan's width. Well over one hundred city blocks would vaporize, or disappear into a hole a quarter of a mile deep, and the effects wouldn't stop there. The almost instantaneous transfer of this incredible amount of energy would create a fireball a mile across and hotter than the sun's surface. The fireball would radiate enough heat to ignite trees and clothing and give people third-degree burns across Manhattan and into New Jersey, Brooklyn, and Queens.

The impact would also generate an atmospheric shock, a blast wave producing high pressures and fierce winds. Within two to three miles from the point of impact—again extending well out of Manhattan into neighboring areas—the blast would severely damage steel-framed buildings and demolish less solid ones, collapse bridges, and overturn vehicles. Even in buildings left standing, every window would shatter, producing hazardous flying glass. The impact would also generate a ground shock, an earthquake of Richter magnitude 5.4—not a major earthquake, but capable of damaging structures within two to three miles.

These horrific effects are like those that would arrive with an H-bomb, with one merciful exception: an asteroid collision does not release deadly nuclear radiation. In Hiroshima and Nagasaki, radiation was responsible for much additional long-term death and injury (as I discuss in a later chapter). Even without radiation effects, though, really big asteroids or comets could threaten life globally, not just over a radius of a few miles.

That kind of real-life event is thought to have wiped out the dinosaurs. In 1980, physicist Luis W. Alvarez and coworkers (including his son Walter) reported finding a geological layer 65 million years old that was unexpectedly rich in the element iridium, which is much rarer on Earth than elsewhere in the solar system. The researchers suggested that a large asteroid had hit the Earth and kicked up a huge amount of dust, made partly of its own material, including iridium, which then settled around the world. This idea was confirmed by later findings of similar iridium deposits in other widely scattered locations.

The time span of 65 million years is significant because that's how long ago dinosaurs disappeared from the Earth, as is known from independent evidence. The hypothesis that explains this timing is that the dust lofted into the atmosphere by the impact blocked out sunlight and remained suspended in the air. With reduced sunshine and significant cooling of the Earth's surface over several years, plant life died. This would have killed species further up the food chain, such as dinosaurs, that depended on plants.

In 1994, fourteen years after the original Alvarez hypothesis, scientists found further supporting evidence: the Chicxulub Crater, 110 miles in diameter, hidden under Mexico's Yucatan Peninsula. Radioactive dating confirmed that the crater was made 65 million years ago. Then in 1998, a piece of meteoritic debris of that same age was found on the floor of the Pacific Ocean, apparently a stray piece of the offending asteroid.

A crater 110 miles across could have been made by an asteroid 6 miles across approaching the Earth at 44,000 mph and delivering 50 million megatons on impact—like a million of the biggest H-bombs ever made, all going off at once. This asteroid hypothesis is a likely explanation for the demise of the dinosaurs, although there remain questions about whether their death was caused by dust alone or by other impact effects as well. For instance, it is not known exactly how long massive amounts of dust would remain suspended in air. Still, that six-mile asteroid is the reason that *Deep Impact* can claim, justifiably, that a such body could wipe out much of life on Earth, though probably not to the extent of killing every microbe.

The description of what a smaller asteroid would do to New York demonstrates that the military use of space projectiles, as shown in *Starship Troopers*, isn't complete fantasy either, if the missile could be accurately aimed. But as presented in *Starship Troopers*, that seems impossible. The Bugs are said to launch the rock that wipes out Buenos Aires from their home planet Klendathu. We're not told how far that is from Earth, but it must be at least four light-years, or 24 trillion miles, the distance to Alpha Centauri, the closest star other than our sun. Apart from the fact that even at the speed of light, four years would pass before the missile reached its target, it's highly unlikely that the Bugs could aim with enough accuracy to hit Earth, let alone a specific city, from that great a distance.

However, it's a different story for missiles launched closer to Earth. From experience with artificial satellites and spacecraft, we know how to aim projectiles from Earth orbit or the Moon with high precision. Robert Heinlein, whose book of the same name inspired *Starship Troopers*, explored this idea in his 1966 story "The Moon Is a Harsh Mistress," in which Moon-based revolutionaries bomb Earth into submission with chunks of moon rock. Real-life versions of such attacks are now under consideration. As part of an effort to build a military presence in space, the U.S. Air Force is studying a proposal called "Rods from God," in which metal rods would be launched from Earth orbit against targets on Earth. The rods, twenty feet long and a foot in diameter, would be made of tungsten or uranium, extraordinarily dense metals that would give each rod considerable heft. Arriving on target at 9,000 mph, each would carry the power of a small tactical nuclear weapon, enough to crack open hardened underground emplacements.

"Rods from God" would be directed against specific targets, but naturally occurring incoming objects land randomly; with the Earth's surface 70 percent water, seven out of every ten objects would hit an ocean, not land. An ocean impact eliminates any chance of smashing into a population center and so might seem less damaging than a land impact, but this is not necessarily so. As the movies show, a large body crashing into an ocean would raise a huge tsunami. Since many low-lying coastal areas are densely populated, this could be extremely destructive to life and property, as the world learned with the great Asian tsunami of December 2004.

If you've ever thrown a pebble into a pond and watched the ripples, you know how impact tsunamis happen. Scaled up to an asteroid- or comet-sized pebble, the crucial questions are: how big is the splash at impact, what happens to the tsunami waves as they travel away from that point, and how do tsunamis behave as they hit a coast and move inland? Conclusions from different computer simulations vary because, without data from observed events—no major impact has ever been observed at sea—scientists don't know exactly how to determine the size of the initial splash and the subsequent behavior. Data from the 2004 Asian tsunami do not help much because that was caused by a undersea earthquake, which doesn't make a tsunami in the same way as a hurtling projectile. Another complication is that tsunamis behave differently in deep and shallow water. As they approach a coast where the seabed shelves upward, they grow in height like any wave at the seashore.

Nevertheless, scientists have developed scenarios based on the best available understanding. One detailed computer simulation, by Steven Ward and

Erik Asphaug of the University of California–Santa Cruz, assumes that 1950 DA (the asteroid with a slight chance of hitting us in 2880), lands in the Atlantic Ocean 380 miles east of Cape Hatteras. In the simulation, the asteroid, delivering 60,000 megatons of energy, blows a cavity in the ocean twelve miles across and deeper than the three-mile ocean depth, uprooting the seabed and making huge waves.

In two hours, tsunamis over 400 feet tall reach the Atlantic seaboard from Cape Cod to Cape Hatteras, typically penetrating inland two to three miles. Four hours after impact, the entire U.S. east coast has been hit by waves 200 feet high or more, and waves at least 80 feet tall reach coastal areas in the Caribbean and parts of South America. Eight hours after impact, waves 50 to 70 feet high reach Europe. Since even tsunamis 30 feet high can do considerable damage, these results show that a body just 0.7 mile across would have extremely serious effects.

Deep Impact envisions an ocean impact from an even bigger object, the 1.5 mile piece of comet that lands in the Atlantic Ocean, also east of Cape Hatteras. In the movie, this creates a tsunami several thousand feet high that crashes into the East Coast of the United States, and creates flooding as far inland as the Appalachians. At least one scientific simulation reproduces this result, but only with a denser and bigger body, an asteroid rather than a comet 3 miles across. This illustrates the uncertainty in these predictions as well as how films may exaggerate the already sufficiently awful effects of an asteroid or comet strike.

Nevertheless, most simulations agree in general scale: an object 0.6 mile (1 kilometer) across would have widespread effects on an entire ocean basin and its coastal areas, and the splash from a six-mile "dinosaur killer" would be horrendous. One calculation predicts that the initial wall of water from such an asteroid would be many miles high, and it would still rear up nearly a mile even 600 miles from the point of impact.

Although the predictions aren't utterly reliable, there's no doubt about the power of a direct collision. However, in *When Worlds Collide*, the planet Zyra doesn't actually hit the Earth but still causes huge disturbances on land and sea as it brushes by. This isn't a wholly imaginary result; it could happen through the power of gravity, which, acting as it does between any two bodies in the universe, produces so-called tidal forces that can tear a planet apart.

Tidal forces can be illustrated by picturing how the Moon's gravity pulls the Earth. Gravity gets weaker with distance, so the pull is strongest on the near side of the Earth and weakest on the far side. This causes the ocean tides and affects the planet itself: it distorts the Earth into a slight egg shape, with two bulges pointing toward and away from the Moon, and exerts a wrenching effect throughout the Earth. Similarly, tidal forces from Jupiter and Saturn constantly flex their moons Europa and Enceladus, warming them sufficiently to maintain water in the liquid state, as I discussed in the last chapter.

In *When Worlds Collide*, Zyra is similar in size and composition to our own planet, making it eighty times as massive as the Moon and giving it an enormous tidal force. If Zyra came anywhere near the Earth, say within a couple of million miles, it would dramatically affect our oceans and cause earthquakes and volcanoes, just as the film shows. And if Zyra came even closer, its tidal force could shatter our world without the two bodies actually colliding.

The distance within which one body can break up another is called the Roche limit, after the nineteenth-century French mathematician who first worked out the possibility. Long ago, the breakup of bodies within the Roche limit caused Saturn's rings. In 1992, comet Shoemaker-Levy 9 came within Jupiter's Roche limit and was pulled apart into the chunks that hit Jupiter two years later. To pull the Earth apart, Zyra would have to approach within about 10,000 miles—in cosmic terms, a thin coat of paint, an incredibly close scrape during which Zyra would loom huge in our skies. *When Worlds Collide* doesn't bring Zyra that close, settling for more limited destruction, but it misses an important fact: while Zyra is tearing at our planet, the Earth is tearing at Zyra. That planet would be subject to similar cataclysms, a terrible inconvenience for the humans who land on it seeking safe haven.

Another piece of science that collision movies don't get completely right is that using a nuclear weapon to break up an incoming object could be a poor approach. The blast may not be powerful enough to do the job, or may not push the pieces into safe trajectories. The pieces may keep right on coming with multiple destructive impacts, as in *Deep Impact*. A better strategy might be to find a way to nudge the intact object off its path. In fact, as we'll see in a later chapter, *Armageddon* has it wrong: even a hydrogen bomb couldn't save the Earth from an object the size of Texas.

Sooner or later, these issues may well become pressing ones for humanity because, despite lapses, the collision films are conceptually correct. Comets and asteroids are out there, and they're potentially a real danger. Although we can't predict their effects with complete accuracy, there's no doubt that a hit by a large object would rank high among natural disasters that afflict our planet. But one unanswered question remains: How often can we expect a devastating collision to happen?

The numbers are favorable in one way: the bigger the object, the smaller the chance it will show up on collision course with the Earth. There are many more small objects than big ones, which means they come more often. Scientists estimate that the Earth suffers one impact every few thousand years from an object 150 to 300 feet across, which as we've seen could have serious effects if it hits exactly the wrong place—but that's unlikely. Big impacts are much rarer: about 200,000 to 300,000 years would elapse between strikes from objects 0.6 mile in diameter, and 60 to 100 million years between impacts from globally life-threatening objects 6 miles across.

Not to discount the human cost of smaller strikes, but the long interval between truly huge strikes shows exactly where it is that collision movies exaggerate the dangers. The films do well in demonstrating the frightening destructive energy carried by a comet or asteroid but less well in conveying the odds that an impact will occur someplace relatively harmless, or so far in the future that we could prepare for it, rather than needing to hastily scrape together Bruce Willis's drilling crew as our sole chance of survival.

The problem, however, is that while the big events are rare, they're also utterly shattering. As Ward and Asphaug write about the possible impact of asteroid 1950 DA, "Humanity lives with a calculus of infinite devastation times infinitesimal probability." Infinite devastation—the obliteration of a city, the raising of unimaginable globe-sweeping mountains of water, the destruction of a planet's life and the planet itself—makes for exciting films. And at the back of our minds is one troublesome fact that hooks us in deep as we watch the films: the probabilities may be low, but they're not zero.

Beyond the mere calculation of what *might* happen to our planet, there's the reality of what *has* happened, and will happen again and again: the destructive impact of the Earth's dynamic natural processes, which I examine next.

Chapter 4

Our Violent Planet

TV ANNOUNCER: [This program] is brought to you by . . . new, delicious, Soylent Green. The miracle food . . . gathered from the oceans of the world.

—*Soylent Green* (1973)

JO [AS A COW IS PICKED UP BY A TORNADO AND IS WHIPPED IN FRONT OF THE TRUCK]: Cow. [Pause] 'Nother cow.
Bill: Actually I think that was the same one.

—*Twister* (1996)

PROFESSOR RAPSON: You remember saying . . . about how melting of the polar ice can disrupt the North Atlantic current?
JACK HALL: Yes.
PROFESSOR RAPSON: Well . . . I think it's happening.

—*The Day After Tomorrow* (2004)

A collision with a giant asteroid or comet might be the worst thing that could happen to us and the other living things on our planet, but that isn't expected to occur very often. Although such a collision probably erased the dinosaurs, we have no evidence that space debris has ever wiped out large numbers of people: one unlucky individual, in 1954, and perhaps a few others who happened to be in the wrong place have ever been hit.

In contrast, our own world generates and supports varied disasters that have killed millions over the centuries and have nothing to do with objects from space but reflect the natural processes of a dynamic planet. The complex interactions among our atmosphere and its clouds, our land and oceans, and the energy from the Sun define our climate and weather. Climatic changes on Earth are long term, such as the cycles that lead to major ice ages. On top of that, we have violent short-term processes: our active, swirling blanket of air supports extreme storms and tornadoes, and our restless seas give birth to hurricanes and carry killer waves.

The Earth is geologically active as well, which creates other destructive forces. Our planet's thin crust is a mosaic of about a dozen tectonic plates, enormous slabs of rock thousands of miles across. These are like rafts floating on a sea of hot magma—that is, molten rock—in the region immediately below, called the mantle. When these moving plates collide with one another, the results are earthquakes, tsunamis, and volcanoes that spew lava, which is magma that has reached the surface. Below the mantle lies another dynamic element, the Earth's even hotter rotating core, part solid and part molten. The iron it contains produces our magnetic field.

Some catastrophes arising from the Earth's built-in dynamics are deadly on large but not global scales. In 1974, in the worst tornado superstorm in U.S. history, 150 twisters killed more than 300 people in thirteen states; in 2004, four hurricanes pelted Caribbean islands and Florida, killing 1,700 people and inflicting $40 billion in damage; and in 2005, Hurricane Katrina battered the U.S. Gulf Coast, led to the flooding of New Orleans, and killed about 1,800 people.

Some natural disasters affect larger regions. The great Lisbon earthquake of 1755 killed 100,000 people and was felt throughout Europe and beyond. The enormous tsunami generated in December 2004 by an undersea earthquake off Sumatra caused 300,000 deaths around the Indian Ocean. This quake had global consequences as well, raising waves that reached the Arctic and Antarctic and the east and west coasts of the United States. Months later, the whole Earth was still vibrating from the seismic energy that was released.

Volcanoes can also affect the whole world. An eruption can spew enough ash and gas into the air to reduce the amount of sunlight reaching the Earth, like a smaller version of a dinosaur-killing asteroid impact. One of the biggest volcanic events ever recorded, the 1883 eruption of Krakatoa, also near Sumatra, killed 36,000 people and put enough material into the atmosphere to cool the world by two degrees Fahrenheit (1.2 Celsius). In 1991, Mount Pinatubo in the Philippines ejected 20 million tons of sulfur dioxide gas (SO_2). This became a mist of sulfuric acid droplets that quickly spread around the world, blocking sunlight and lowering global temperatures by one degree Fahrenheit.

Another worldwide temperature and climate change now underway—a potential catastrophe in the making—probably part natural and in large part human-made, is unfolding more slowly than a volcanic eruption. This is global warming, the gradual increase of the Earth's temperature that scientists have measured over the last century.

Our planet's temperature is affected by the sunlight it absorbs. After sunlight heats the Earth's surface, the heated areas produce invisible infrared radiation that works its way back up through the atmosphere. But some outbound energy is reflected back to the surface rather than escaping to outer space. This is the greenhouse effect, where sunlight shining through glass creates heat that can't follow the reverse path back through the glass and so provides a warm environment. Similarly, radiation caught between the Earth's surface and its atmosphere heats the Earth. Instead of glass, certain greenhouse gases in our atmosphere, especially carbon dioxide (CO_2), trap the heat.

An extreme greenhouse effect is a large part of the reason the planet Venus is the hellhole that it is. Its atmosphere is 96 percent CO_2, trapping heat so efficiently that the surface temperature approaches 900 degrees Fahrenheit. The Earthly concentration of CO_2 is far lower, only parts per million (we couldn't survive otherwise), but there is concern that human activities, mainly the burning of fossil fuels, are increasing the concentration and accelerating global warming.

Automobiles, coal-burning power plants, and other sources emit billions of tons of CO_2 and other greenhouse gases yearly. No one thinks this will turn the Earth into another Venus, but the evidence indicates that these activities are raising global temperatures. One result would be the melting of the ice at the North and South Poles, foreshadowed by the retreating glaciers already seen around the world. Along with the expansion of sea water as it warms, this melting would raise ocean levels and cause flooding (except in the case of the portion of Arctic ice that floats like ice cubes in a drink; when it melts, water replaces the ice without raising ocean levels). Other possible outcomes, with varying probabilities, include violent weather, effects on ecology and agriculture, and, paradoxically, a chance of a new ice age.

Compared to the excitement and grand scope of facing aliens or colliding with an asteroid, most Earth-based disasters may seem small potatoes. Global warming, however, could produce as huge a planetary cataclysm as any movie director could want, but unfortunately from the cinematic viewpoint, it's developing over decades, not minutes and hours like a good hearty disaster. As the main event of a two-hour film, it would be as thrilling as watching grass grow. One cinematic approach to such a lack of drama is to enhance nature by exaggerating or speeding up the destructive process. Another is to bypass the catastrophe itself and show only its aftermath, its effects on us and our world.

However, some of our planetary violence is so cinematic that it automatically makes for gripping movie scenes, even when it's not global. For example, in *Twister* (1996), Dr. JoAnne "Jo" Thornton-Harding (Helen Hunt), whose father was killed by a tornado when she was a child, thrives on an intense cocktail of adrenaline and science as she chases these killer storms across Oklahoma. She and her crew tool around the countryside in a motley collection of vehicles outfitted with weather instruments. Whooping with joy and listening to hard rock, they do their level best to get near tornadoes so they can analyze them.

On the day that Jo's husband and fellow tornado hunter Bill Harding (Bill Paxton) shows up to get his divorce papers, Jo's ready to try a new instrument. Named "Dorothy" after the world's most famous tornado rider in *The Wizard of Oz*, it can launch hundreds of instrument globes the size of Christmas tree ornaments into a tornado. The sensors will report wind speed and so on throughout the funnel, where no data have ever been gathered—knowledge that will save lives by giving better predictions. But to get Dorothy into the funnel, the hunters have to set the device in the tornado's destructive path where the sensors will be swept aloft, and then pray they can get out of the way in time.

Bill had planned to leave both Jo and the tornado business, remarry, and start a quieter life, but he is pulled back into the excitement of the chase. He's competing, too, with other tornado hunters under Dr. Jonas Miller (Cary Elwes), who, according to Bill, has great equipment but lacks good tornado instincts. Fortunately, Bill's nose for the erratic behavior of tornadoes brings him and Jo close to several—dangerously so. They brave flying debris, from a motorboat to a cow, and ride out one twister under a wooden bridge as the wind plucks out its very nails, but they have no luck in launching Dorothy's sensors.

Finally, they get word of a level F5 storm, tops on the scale of tornado violence, with winds of 300 mph. As Jo and Bill, and Jonas in his own vehicle, approach this mile-wide monster, Jonas guesses wrong about its path. A flying television tower impales his SUV, and the whole vehicle is picked up, whirled around, and dropped to the ground where it explodes.

Jo and Bill have their own problems. They drive straight through a really big piece of debris, a house the twister drops on the road, and risk their lives to put Dorothy in the tornado's path. The sensors immediately whirl up and start sending data to Jo's madly cheering crew. But the tornado veers toward Jo and Bill, and they run for shelter in a shed that disintegrates as they lash themselves to water pipes. The wind blows their bodies straight out like flags and then upside down, so for one eerie moment, they look right up into the funnel. The pipes hold, however, the tornado dissipates, and though surrounded by wreckage, Jo and Bill are fine. Excited and reunited by their

Twister (1996). After barely managing to place recording devices into a violent level F5 tornado, married tornado experts Bill Harding (Bill Paxton) and JoAnne "Jo" Thornton-Harding (Helen Hunt) frantically outrun the storm as it rips up the fence behind them. Tornadoes of this strength can actually move cars and buildings, as the film shows (and apparently can also heal shaky marriages; in the film, their shared danger brings Bill and Jo back together).

Source: WB/Universal/Amblin/The Kobal Collection.

breakthrough, they won't divorce after all; they kiss, and squabble happily over who'll run the tornado lab and who'll analyze their new data.

Krakatoa, East of Java (1969), *Dante's Peak* (1997), and *Volcano* (1997) also use dramatic cinematic moments while conveying something about volcanoes. In *Volcano*, workaholic Mike Roark (Tommy Lee Jones), who runs the Los Angeles Office of Emergency Management, has dealt with disasters from mudslides to earthquakes. But then he sees signs of a different kind of event. The La Brea tar pits in the city's heart are unusually active. Nearby, city workers are found dead in a storm drain. Their deaths are chalked up to scalding from a broken steam pipe, but an attending physician notices that the victim's clothes are charred as if by flame.

Mike finds cracks in the drain and feels extreme heat even through protective gear. Something unusual is happening, he decides, and he shouts "Find me a scientist!" as he climbs back to the street. The scientist he gets is ultra-self-possessed geologist Dr. Amy Barnes (Anne Heche). She finds that the temperature of a nearby lake has risen six degrees Fahrenheit in just twelve hours, a sign of major geological doings.

Amy suspects that a volcano may be forming. She and her colleague Rachel (Laurie Lathem) tell Roark that in 1943, a volcano suddenly grew in a Mexican cornfield without prior warning. To confirm her suspicion, and despite the danger, Amy climbs down into the storm drain with Rachel. As they find evidence of sulfur, a dead giveaway for volcanic activity, Los Angeles is hit by a big quake centered in the Mojave Desert. A chasm opens up in the drain, and although Amy tries to save her, Rachel falls in.

Amy escapes, while Roark, driving to OEM headquarters with his thirteen-year-old daughter, Kelly (Gaby Hoffmann), sees a huge glowing plume erupt out of the La Brea pits. Hot chunks of debris shoot out like meteors that threaten people and wreck buildings and vehicles; ash drifts down like surreal gray snow; and a stream of intensely hot, evil-looking lava flows down a main thoroughfare, Wilshire Boulevard, frying or melting everything it touches.

Mike and Amy work frantically to help people on the scene, including Kelly, who's been burned, and to halt the lava flow. Mike gets a bus overturned to form a barrier between the lava and the Los Angeles art museum. Then, as Amy suggests, he dams up the lava flow with concrete road barriers and cools

Volcano (1997). Fire engines frantically pump cooling water onto a lava stream advancing down Wilshire Boulevard after a volcano unexpectedly emerges in the heart of Los Angeles. Although a volcano in Southern California is geologically unlikely, this scene follows an actual strategy once used to stop a lava flow in Iceland.

Source: 20th Century Fox/The Kobal Collection.

it with a barrage of water from fire engines and helicopters until the lava solidifies, removing the threat to people's homes.

But Mike and Amy find more lava beneath the city, ready to emerge at a hospital filled with thousands of volcano victims, including Kelly. Mike comes up with an audacious plan: bring down a building to protect the hospital from this new danger and divert the lava into a nearby stream that will carry it harmlessly into the Pacific Ocean. An army of demolitions experts sets explosives, but as these go off, Mike sees Kelly and a little boy she's been babysitting. He runs toward them as the building collapses perfectly to save the hospital and send the lava to the sea. Though buried in rubble, Mike, Kelly, and the child emerge unharmed. As a cleansing rain washes the scene, we see that Mike and his daughter have been brought closer, and maybe Mike and Amy, too. And as for Los Angeles? It's now the proud possessor of an active volcano, Mt. Wilshire.

Dante's Peak (1997) also tells the story of a volcanic eruption. Dante's Peak, voted the second-best U.S. small town in which to live, is nestled on the slopes of an extinct volcano in the Northern Cascades, in Washington state—except that according to volcanologist Dr. Harry Dalton (Pierce Brosnan) of the U.S. Geological Survey, volcanoes are never extinct, only dormant. Harry should know; he has traveled everyplace in the world where there's a volcano with an attitude, even losing the woman he loved to an eruption in Colombia.

Harry is in Dante's Peak to check on signs of volcanic unrest. Attractive single mom Rachel Wando (Linda Hamilton), mayor and coffee shop proprietor, tags along with her two kids. Harry soon spots dead trees and squirrels and measures high acidity in the local lake. Worst of all, the group comes upon two bathers scalded to death in a previously safe hot spring. All this points to one fact—the volcano is about to blow. But Harry meets stiff resistance when he asks the town council to consider evacuation. The town has just lured in new investment and jobs, and the council is reluctant to upset the applecart.

Harry's boss, Dr. Paul Dreyfus (Charles Hallahan), backs the town council. He's cautious about crying wolf but promises that his USGS team will monitor the volcano and warn the town if there's a problem. Using a walking robot called Spiderlegs, the team investigates the huge volcanic crater. There are no signs of dangerous activity, and Paul remains unconvinced that something is brewing.

But Harry finally gets proof that it's really serious when he finds that the town's water has turned brown and smells of sulfur. As he gives the bad news to a packed town meeting, the building starts shaking, and the volcano emits a huge cloud. Panic sets in, and people stampede as they frantically try to escape. Vehicles smash into one another and roads become impassable while the church steeple goes down. A helicopter crashes as it lifts off, its turbines clogged by the thick fall of ash.

Harry ends up rescuing Rachel's family including her former mother-in-law, Ruth (Elizabeth Hoffman), from Ruth's cabin high on the volcano. They escape the flowing lava in a boat, but the acid water eats the outboard motor and Ruth's legs so badly that she dies. Harry gets everyone else, including the family dog, into a pickup truck as even worse news comes: the volcano is generating a pyroclastic flow, an avalanche of immensely hot gas, rock, and ash rushing down the volcano's side faster than a speeding car, mowing down trees and buildings at an incredible rate. Harry escapes it only by ramming the truck into an abandoned mine shaft.

Trapped in the shaft after its roof caves in, there's only one hope: try to signal rescuers with a special radio Harry had thoughtfully remembered to stow in the truck. Although Harry's leg is broken and he's pinned down with barely room to breathe, he manages to flip on the transmitter and contact the USGS team. Soon heavy equipment is digging everyone out of the shaft, not forgetting the dog. There are hugs all around, and Harry promises to take the family on a fishing trip as he and Rachel hold hands in the rescue helicopter.

The Core (2003) does more than show lava; it takes us deep into the Earth through the magma to the core, site of a major disaster, although its first signs are puzzling: thirty-two people with cardiac pacemakers suddenly drop dead. General Thomas Purcell (Richard Jenkins) asks geophysicist Dr. Josh Keyes (Aaron Eckhart) and weapons expert Dr. Serge Leveque (Tchéky Karyo) if this could have been caused by an electromagnetic weapon. They think not, but meanwhile further strangeness happens. Flocks of pigeons in Trafalgar Square fall from the sky to smash into people and windows. In Washington, D.C., people are astonished to see the aurora borealis, the northern lights, so far south.

Josh determines that something is wrong with the Earth's core and tells eminent scientist Dr. Conrad Zimsky (Stanley Tucci), who then has Josh summoned to the Pentagon. Using a peach as a prop, Josh shows the assembled military brass what's going on. The skin, meat, and pit correspond to the Earth's crust, mantle, and finally the core, which has stopped rotating for reasons unknown. That turns off our "electromagnetic field of energy," affecting electronic technology and exposing the Earth to microwaves that will cook us in a year, as Josh illustrates by burning the peach to a cinder. General Purcell asks "What can we do?" Josh points out that the deepest we've ever drilled is seven miles, so there's no way we can reach the core, whose outermost edge is 1,600 miles down.

But we can, as shown by the next scenes. At a laboratory in Utah, Zimsky's former colleague, Dr. Ed "Braz" Brazzleton (Delroy Lindo), has invented a combined laser-ultrasound system that chews through rock like a buzzsaw through cheese. He's also developed "unobtainium," a material with the wondrous properties of becoming stronger under the heat and pressure within the Earth and of turning heat into energy. It would normally take years to build an unobtainium vehicle, but Purcell funds Braz with $50 billion to finish the craft in three months.

The *Virgil* (after the classical Roman poet who wrote about Hades) will carry Josh, Zimsky, Braz, and Serge to the core, where they'll set off 1,000 megatons worth of hydrogen bombs to restart the rotation. Their pilots are two NASA astronauts, chosen because they made a brilliant landing of their off-course space shuttle onto the bed of the Los Angeles River. Another team member is Taz "Rat" Finch (D. J. Qualls), a young genius of a computer hacker who'll work at Deep Earth, *Virgil*'s surface control center, to keep the project secret.

Virgil is completed and launched, downward naturally, into the Pacific Ocean and then the crust. Cruising at 60 knots, everything works well until, 700 miles down, a huge crystal in a giant geode jams *Virgil*'s laser. The terranauts cut away the crystal, but lose one pilot to a pool of hot magma, leaving only Major Rebecca "Bec" Childs (Hilary Swank). Then Serge dies after a diamond the size of Cape Cod breaches the hull.

More trouble looms when it turns out that 1,000 megatons won't be enough. Zimsky advises giving up on the mission, and sends a message to General Purcell saying "DESTINI is a go." This is the Deep Earth Seismic Trigger Initiative. It's a secret weapon that can initiate earthquakes, and in fact, it's what stopped the core's spin when the military tested it. Purcell gives the order to fire up the device, hoping to restore the rotation. But over Zimsky's hysterical protests, the *Virgil*'s crew vows to continue downward and set off their bombs before DESTINI causes further damage.

Meanwhile, conditions on the surface are dire. The Golden Gate Bridge is hit with so much microwave energy that its cables sag and the bridge collapses, throwing people and cars into the bay, and lightning destroys the Roman Colosseum (also, blessedly, lightning nails that Roman eyesore, the Victor Emanuel monument).

When Zimsky stops sulking, he calculates how to explode the five 200-megaton bombs in a reinforcing pattern that will restart the core. To set up the bombs, Braz braves the outside temperature of 9,000 degrees Fahrenheit, and lasts long enough to do the job. But the scientists find yet another glitch: the last bomb isn't quite powerful enough. Unluckily, Zimsky is trapped in an inaccessible compartment, but before he dies, tells Josh the solution: use the fuel rods from *Virgil*'s nuclear reactor.

Josh adds the rods, the explosions go off perfectly, and Deep Earth reports that the core has full rotation. Our planet is saved, but Josh and Bec are stuck, out of power, until Josh remembers that unobtainium converts heat to energy. That allows *Virgil* to climb back to the sea bottom near Hawaii, where Josh and Bec are retrieved. In the final scenes, Rat puts the whole secret story out over the Internet, to honor the living and dead terranauts.

More likely than having our core screech to a halt is the possibility of global warming. It has become so much part of the film lexicon that it can be a plot element even when it's not the main story. Steven Spielberg's *A.I.: Artificial Intelligence* (2001) begins with an image of the ocean, as we hear in voiceover that the ecology has gone bad, the polar ice caps have melted, coastal cities are underwater, and there's not enough food. That motivates this story of a future world where childbearing is limited and where robotics expert Professor Allen Hobby (William Hurt) makes a robot child that can love, as I discuss in a later chapter.

Global warming also appears in *Soylent Green* (1973), which shows directly what can happen as a result of disregard for the environment and population growth. The movie is set in the year 2022, with humanity in a bad way. A series of still photos shows why: starting with sepia-toned nineteenth-century photographs, we see an ever more crowded world until, by 2022, the population of New York City is 40 million (with 20 million unemployed), five times what it is now.

Along with population growth, the Earth's capacity to support people has actually been reduced. Through war and nasty industrial byproducts, humanity has poisoned the water, soil, and air and devastated plant and animal life. This abuse has also accelerated global warming, and it's hot. As Robert Thorn (Charlton Heston) says, only with air conditioning does it ever get as cold as winter used to be.

Thorn is a detective with the New York Police Department. As he goes through his day, we see how impoverished and crowded the world has become. Breakfast with his elderly partner, Sol Roth (Edward G. Robinson, in his last film) consists of "tasteless, odorless crud," says Sol. Natural food such as beef and strawberries is an expensive rarity. Half the world's food supply is in the form of Soylent, a concoction that comes in three colors: Red and

Yellow, made from soybeans and lentils; and Soylent Green, which, according to its advertising, uses plankton harvested from the oceans, and is so popular it can't be kept in stock.

Leaving his apartment, Thorn climbs over dozens of people sleeping on the stairs. Outside, hordes of drably dressed, poorly fed citizens roam the dusty streets while waiting for their allotment of food and water. New York resembles a struggling third-world city, not a great metropolis. At police headquarters, Thorn is sent to investigate the murder of William Simonson (Joseph Cotten), whose head has been bashed in with a crowbar, presumably by a burglar. But from earlier scenes, we know that Simonson was actually assassinated by a hit man whose weapon was supplied by State Security Chief Donovan (Roy Jenson), and that Simonson seemed almost to accept his death.

Thorn knows nothing of this but sees that Simonson lived in an elegant, luxurious apartment with hot water and a shower, air conditioning, and plenty of food and drink, including real beef. He interviews Simonson's female companion and his bodyguard. Then he leaves, after helping himself to fresh food, a flask of bourbon, and, for his scholarly partner Sol, a volume labeled *Soylent Oceanographic Survey Report, 2015–2019* (in this diminished society, paper and printing presses are in short supply, and books are uncommon).

Thorn finds that Simonson sat on the board of directors for the Soylent company and that he was extremely troubled. Simonson visited a Catholic church just before he died so he could confess to a priest, but other than hinting that Simonson's secret was truly awful, the priest tells Thorn nothing. It's Sol and his bookish friends, examining the *Oceanographic Survey Report*, who find the secret that Simonson had also uncovered. It's so ghastly that it shook Simonson's sanity, and the corporation eliminated him because he had become unreliable.

For Sol, the secret is equally unendurable and he leaves a note, "I'm going home," which sends Thorn rushing to the government euthanasia center. There Sol has agreed to be put to death, and solicitous attendants have given him a death potion. As Sol watches beautiful scenes of unspoiled nature, from flowers to deer, oceans, mountains, and birds—that is, the world as it used to be—and listens to classical music, he slowly and happily sinks into oblivion, but Thorn insists on speaking to him before he's gone. Sol says goodbye, then becomes agitated and blurts out "listen . . . horrible . . . Simonson . . . Soylent . . . you've got to prove it," and dies.

Thorn makes his way to the hidden part of the euthanasia operation, where the bodies, in white shrouds, are loaded off white gurneys and stacked in garbage trucks. Climbing onto a truck, he ends up in a heavily guarded compound. There Thorn stealthily enters a huge factory and follows the bodies as they're dumped onto a conveyer belt. He finally penetrates to the heart of the secret when he sees another conveyer belt filled with wafers of Soylent Green, and he realizes that the food coming out is made from the bodies going in.

Thorn escapes back to the city, but he's attacked by government and corporate forces who want to preserve the secret. He ends up with a gunshot wound in the same church where Simonson made his confession. As he's carried out on a stretcher, Thorn gathers his strength and shouts,

> You've got to tell them . . . I've got proof . . . I've seen it happening . . . the oceans are dying, the plankton's dying . . . it's people . . . Soylent Green is made out of people . . . next thing they'll be breeding us like cattle for food . . . you've got to tell them . . . Soylent Green is people . . . we've got to stop them somehow . . .

And the movie ends on this unresolved and hopeless note.

Waterworld (1995) also shows a future world after global warming. As in *A.I.*, the polar ice has melted, but this time, the whole world is covered with water. People live in floating towns called atolls, or like the Mariner (Kevin Costner), crisscross the limitless ocean. When we first meet the Mariner on his big sailing trimaran, it's in a decidedly unheroic way: he urinates into a bottle, recycles the liquid through a still, and drinks the result, with some left over to water a little lime tree in a pot. This is a world where fresh water, plain dirt, and growing plants are precious commodities, and where people believe in the semimythical Dryland as a Shangri-La.

The Mariner is held and imprisoned on an atoll partly because he's a mutant, with gills as well as lungs, and webbed feet. When Hells Angels–like pirates called Smokers attack the atoll using jet skis, motorboats, and an aircraft, the Mariner escapes on his boat with shopkeeper Helen (Jeanne Tripplehorn) and her young ward Enola (Tina Majorino). Tattooed on Enola's back is a map supposedly showing the way to Dryland, which interests the Smoker's chief, Deacon (Dennis Hopper). He has promised to lead his gang to Dryland and pursues Mariner to get the map.

In a nod to the role of fossil fuels in global warming, and in contrast to the natural life of the Mariner and the Atollers, the Smokers hang out on the rusty hulk of the *Exxon Valdez*—the huge tanker that hit a reef in 1989 and spilled a sizable part of its 50 million gallons of oil, causing a terrible environmental impact on the Alaskan coast. The Smokers use the remaining oil to fuel their attack craft and a vintage convertible they drive around the tanker's vast deck, and they keep a photo of Joseph Hazelwood, the actual captain of the *Valdez* in 1989, in a place of honor.

Deacon captures Enola, but the Mariner rescues her, blows up the *Valdez* by igniting its oil, and escapes with the help of Helen and other Atollers. They follow Enola's map and find that Dryland is no myth: it's real, a paradise of beaches, mountains, jungles, and waterfalls, with animals from birds to horses that escaped the worldwide flood. To the Atollers, this is heaven, and they plan to stay, but the Mariner says it's not for him. He goes to sea again in a new boat, while Helen and Enola sadly watch him leave.

Global warming reaches fever pitch, so to speak, in *The Day After Tomorrow* (2004). Climatologist Jack Hall (Dennis Quaid) is drilling ice cores on the Antarctic ice shelf to examine trapped greenhouse gases. Suddenly, the ice cracks and develops huge crevasses. Jack leaps across one to save his precious cores, but doesn't quite make the leap back and is barely saved by his colleagues, one of whom says "the whole damn shelf is breaking off."

This is a sign of the coming climatic change indicated by Jack's research, as he reports to a UN conference in New Delhi. Against intuition, he explains, global warming can cause a new ice age. As the temperature rise melts the polar icecaps, fresh water will disrupt the ocean current that carries heat from the Equator to the Northern Hemisphere. The burning of fossil fuels and environmental pollution are speeding up this process.

The political implications of Jack's research become clear when U.S. vice president Becker (Kenneth Welsh, who resembles Dick Cheney) asks who'll pay the hundreds of billions of dollars needed to implement the Kyoto Accord, the international agreement to reduce greenhouse gases. Jack replies: "With all due respect, Mr. Vice President, the cost of doing nothing could be even higher." But Becker accuses Jack of "sensationalistic" claims.

Nevertheless, something is stirring, for when Jack leaves the conference, it's snowing, the first time ever in New Delhi. Other signs include an abrupt drop in ocean temperatures reported by Professor Terry Rapson's (Ian Holm) team in Scotland, huge hailstones in Tokyo, and record-setting hurricanes. Back at NOAA (the National Atmospheric and Oceanic Administration) in Washington, D.C., Jack works to predict the coming changes, while dealing with his wife (Sela Ward) and teenage son Sam (Jake Gyllenhaal). Jack is too driven a researcher to relate well to his family, and he fumbles a commitment to get Sam to the airport so the boy can fly to New York for a competitive scholastic decathlon (Sam, says Jack, knows more science than the vice president).

Meanwhile, unceasing rain falls on New York, tornadoes ravage Los Angeles and shred the "Hollywood" sign, and Rapson confirms that the polar melt is disrupting ocean currents. All travel is suspended, trapping Sam and his friends in New York. With others, they seek refuge in Manhattan's

The Day After Tomorrow (2004). Even hardened, blasé New Yorkers take note of a tsunami inundating the Forty-second Street Library, along with incredible storms and an ice age, all due to global warming and shown through intense special effects. The film exaggerates these phenomena and gets some of them wrong, but it has a core of real science and points to a real issue. *Special Award; see chapter 9.*

Source: 20th Century Fox/The Kobal Collection.

immense Forty-second Street library to escape flooding caused by a tsunami. And the predicted cold is coming. Royal Air Force helicopters flying to rescue the Queen from Balmoral Castle in Scotland crash when extremely low temperatures freeze their fuel lines. But Rapson's scientific team remains on duty, sending data to Jack until they're snowed in beyond hope of rescue.

When Jack finally gets to brief the U.S. president, he explains that three huge hurricane-like storms have formed and will soon combine into a superstorm. In seven to ten days, the Northern Hemisphere will be covered with snow and ice, which strongly reflect sunlight. That will change atmospheric conditions and cause a new ice age.

With the government at last realizing that matters are serious and evacuating millions of Americans to Mexico, Jack dons arctic gear and with two colleagues drives off to rescue his son. Meanwhile, Sam's group keeps warm by burning library books (although they save a Gutenberg bible), but Sam's would-be girlfriend, Laura (Emmy Rossum), comes down with a serious infection. Although Sam and his friends find penicillin aboard a Russian cargo ship that drifted into the city, they have to fight off a ravenous wolf pack before they can make it back to the library.

Finally the superstorm starts to clear, but the scene is still bleak when Jack reaches New York. The frozen harbor is full of ice-bound ships, and massive icicles cover the Statue of Liberty. Jack and his remaining companion continue to the library, which is buried in snow, make their way inside, and find the survivors. The hug between Jack and Sam expresses Jack's commitment to become a better father.

Then Jack informs the U.S. government in Mexico of the successful rescue in New York. The president was lost attempting to leave Washington, and has been replaced by Becker, the vice president, who had rejected Jack's analysis. Now chastened, he says in a broadcast, "We are left with a profound sense of humility in the face of Nature's destructive power. For years we operated under the belief that we could continue consuming our planet's natural resources without consequences. We were wrong. I was wrong."

In the last scenes, a helicopter gathers up the library survivors and swings south toward Mexico as Laura lays her head on Sam's shoulder. The final comment comes as astronauts in their space station gaze down at a beautiful glowing Earth and exclaim at the clarity of its air.

These films begin with basic scientific facts, and some reflect the science pretty well throughout the story. For instance, the issues presented in *Twister* are real. We don't know all that much about tornadoes. Although it's clear that they start from powerful thunderstorms, the details of how they form are imperfectly understood. Predictions of the timing, location, intensity, and path of tornadoes are too inexact to give much warning.

To get better data, NOAA's National Severe Storm Laboratory in Norman, Oklahoma, sent out tornado chase teams in 1994 and 1995. These were better equipped than those in *Twister*, with a dozen vehicles, a sophisticated mobile radar, and two aircraft; the hard rock and cowboy whoops heard in *Twister* were optional. *Twister* may also have been overly colorful in the variety of flying debris it showed and overly optimistic in letting Bill and Jo survive that F5 tornado, but the numerical rating and the destruction are real. They come from a rating system invented by meteorologist Tetsuya Fujita of the University of Chicago, which describes an F5 tornado as causing "incredible damage"; it can lift frame houses off their foundations and carry a vehicle a hundred yards, as shown in the film.

"Dorothy" in the film is based somewhat on a real device, too, with another name out of the *Wizard of Oz*. That was TOTO, for Totable Tornado Observatory, an instrument package designed to be put in the path of a tornado, but meant to stay on the ground; it wasn't packed with aerodynamic sensors like Dorothy. Nor was TOTO ever successfully deployed, so Bill, Jo, and Dorothy go beyond real science in that aspect.

Other films walk a shakier line between real and imaginary science, as illustrated by the "volcano and core" films *Krakatoa, East of Java*; *Dante's Peak*; *Volcano*; and *The Core*. *Krakatoa, East of Java* immediately gets off on the wrong foot, since Krakatoa is actually west of Java, but it's the only film of the four to weave a real event into the story. Captain Hanson (Maximilian Schell), of the East Indies steamer *Batavia Queen*, sets sail to seek a sunken cargo of pearls. Along with problems aboard ship, including vicious convicts who try to take it over, and a diver (Brian Keith) who gets into violent rages, there's the unfortunate fact that the treasure lies near the volcano that's about to erupt. The movie doesn't deliver a great deal of science, but what it shows follows what really happened: warning signs before the eruption, an explosion that blows away much of a whole island, falling debris that bombards an immense area, and huge loss of life as a giant tsunami batters low-lying islands, although

Krakatoa, East of Java (1969). The Krakatoa volcano in Indonesia begins to get violent in this scene from a film that, unlike other disaster movies, shows a real-life catastrophe: the immense Krakatoa eruption of 1883, which had worldwide consequences. The film presents a reasonable sketch of volcanic effects, though its credibility suffers when it relocates the volcano many miles from its true site west of Java.

Source: ABC/Cinerama/The Kobal Collection.

the *Batavia Queen* survives. This is reminiscent of the great 2004 tsunami in the same area, which, however, came from an undersea earthquake, not an eruption.

Dante's Peak is also fairly accurate about the behavior of volcanoes and volcanologists. The kinds of warning signs volcanologists seek and the kind of testing they do are portrayed with some realism; even the science-fictionish robot Spiderlegs that walks into the volcano's crater is modeled on a real robot used by volcanologists. Also accurate is the uncertainty of volcanic prediction, even with the best data. Everything the volcano does in the movie represents real behavior as well, with some exceptions. Northern

Cascade volcanoes don't spew out fast-flowing lava of the type shown, nor would they produce lava and a pyroclastic flow at the same time, and the acid lake that melts outboard motors couldn't happen as shown. Still, one group of geologists rated the scientific content of the film as B+ to A–, a higher rating than the science in many other movies would get.

Volcano also has bits of reality; for instance, when geologist Amy Barnes tells Roark, correctly, that volcanoes happen when magma wells up along the boundaries between tectonic plates. It's also true that in 1943, the volcano Paricutín erupted suddenly in a farmer's field in Mexico, and, in 1973, the inhabitants of a town in Iceland stopped a lava flow by pumping cooling water onto it.

But dig deeper and you see where the science goes wrong. Volcanoes occur where two tectonic plates converge, which means they collide head on, pushing or subducting one plate down under the other and allowing magma to rise; or where they diverge, again opening up a passage for magma. Although Los Angeles lies near the San Andreas fault, a boundary between two tectonic plates, this is a transitional boundary, where the plates don't converge or diverge but slide past each other. That produces problems enough for Californians in the form of earthquakes, but not brand new volcanoes. Paricutín in Mexico happened in a region already known for its volcanoes.

Similarly, rather than the puny streams of water and quick success shown in the film, it took fierce pumping of millions of gallons of seawater, and many days, before the lava in Iceland stopped flowing. Also, if you ever have to flee from a river of lava, don't rely on a bus or concrete barriers to protect you. Lava can be as hot as 2,200 degrees Fahrenheit (1,200 Celsius), enough to soften or melt most materials, including concrete.

Volcano, however, is a feast of science compared to *The Core*. When Josh describes the Earth's internal layers, crust to core, that's the first and last bit of correct science; everything else is pretty much dead wrong, starting when Josh says the core makes an "electromagnetic field of energy." The Earth's core makes a *magnetic* field like that from the magnet on your refrigerator, related to but different from an electromagnetic field, and neither kind of field is synonymous with energy.

Other examples abound: microwave radiation is a tiny part of what's out there in the universe, not enough to melt a bridge or fry the Earth; in any case, microwaves aren't deflected by a magnetic field so it makes no difference whether the field is on or off; it's ludicrous that Braz could survive temperatures like those at the Sun's surface; and if you really wanted to use an H-bomb to make the core spin—not a great idea at best—the explosion would have to push in a specific direction to create a turning effect, whereas an H-bomb explosion spreads out in all directions.

There's more, much more, to say about *The Core*, such as its choice of where to launch *Virgil* and the amazing properties of unobtanium, but let's leave it as the movie that put the fiction into science fiction and turn to films about global warming. Is it really happening, and will it raise temperatures substantially by 2022, as in *Soylent Green*? Is it true that warming could lead to incredibly violent weather and a colder climate, as in *The Day After Tomorrow*, and as rapidly as portrayed there? Are ocean levels rising, and if so, are we talking just wet feet, or a planet submerged as in *Waterworld*?

There are real answers to these questions, although not always definitive ones. Global warming and its effects depend on the Sun's energy output and the Earth's relation to the Sun and on the dynamics and makeup of our atmosphere, the oceans and their currents, and the polar ice caps and glaciers. Boiling down all these processes to define the behavior of a whole world is difficult.

Still, scientists agree about some central facts. Over the last 1,000 years, global temperatures have tracked up and down with natural events like volcanic eruptions, until the twentieth century. Then temperature started rising rapidly, going up about one degree Fahrenheit in the last 100 years. This is connected with the rise in greenhouse gases because of human activities. The concentration of CO_2 has increased some 30 percent between 1750, before the world was industrialized, and now (and is still climbing, with other greenhouse gases). That's why, in 2001, the United Nations Intergovernmental Panel on Climate Change (IPCC), an international group of climate scientists, declared, "There is new and stronger evidence that most of the warming observed over the last 50 years is attributable to human activities . . . [it] is likely to have been due to the increase in greenhouse gas concentrations," and added that most of the greenhouse increase comes from the burning of fossil fuels. Additional data were analyzed in the latest IPCC report, issued in early 2007, which stated that "warming of the climate system is unequivocal." The 2007 report upgraded the impact of human activity on warming from "likely" to "very likely," meaning that scientists are better than 90 percent confident in their assessment of the human role.

Even if we could instantly turn off the human contribution, global warming is like a car that doesn't stop moving the second you turn off the engine: its inertia keeps it going, and thermal inertia guarantees that the warming trend will continue. But it's not easy to forecast just how much global temperatures will increase in coming decades and how that will change the world. In 2004, the consensus among climate researchers was this: if CO_2 concentration were to double, as expected by the end of the century (if we continue doing what we're doing), that would raise global temperatures roughly five degrees Fahrenheit (three degrees Celsius). An increase of five degrees is also presented in the 2007 IPCC report as the most likely value.

That may not seem like much. It's far less than implied in *Soylent Green*, where in 2022, long before century's end, winter no longer exists. But a five-degree warming is enough to suggest those dramatic flooding sequences seen in films because it's projected to raise the sea level up to two feet by 2100, mainly from expansion of seawater, with a further increase coming from the melting of the world's ice. A two foot rise may not seem like much either, but it would eliminate 10,000 square miles in the United States, the combined area of Massachusetts and Delaware. Low-lying cities, like Boston and New York, would flood to some extent, at least during storms. Those below sea level, like New Orleans, would be even more vulnerable—as shown with terrible clarity in 2005, when Hurricane Katrina hit the city and the levees broke—along with low-lying areas elsewhere. These include places hit hard by the great 2004 tsunami, such as Indonesia and Thailand, and parts of the Mediterranean coast. But we won't see waves lapping against the upper stories of New York's skyscrapers, as shown in *A.I.*, any time this century.

The scenario in *Waterworld*, Earth covered by water, is even less likely, in fact, impossible. Take the extreme case: if all the world's ice were to melt, in Antarctica, Greenland, and from every glacier everywhere—which would require nightmarish levels of climate change—the sea would rise well over 200 feet (70 meters). This is a lot, enough to put whole cities underwater, as a scene in *Waterworld* shows, but much of the land would remain unflooded. There just couldn't be enough melted ice to make the whole world one big ocean.

A variety of other climate-related effects is shown or implied in these movies. In *Soylent Green*, it's the dying plankton, which leads to institutionalized cannibalism. Not much is known about the effects of global warming on plankton and marine life in general, although the warming does seem to be damaging coral reefs, which are living systems. But one recent study shows that the abundance of plankton goes down as water temperature goes up. If this behavior is confirmed, it's important because plankton are essential to the food chain that includes edible fish. In that way, global warming could lead directly to a *Soylent Green* world, without enough natural food from land or sea to feed all of humanity.

In *The Day After Tomorrow*, the main effects are distorted weather patterns, such as killer tornadoes in Los Angeles, and the descent of an ice age—all of which happens really fast, in days or weeks. It's easy to find anecdotal evidence for connections between global warming and weather. In the same 2004 season that four hurricanes struck Florida, ten typhoons hit Japan. Both numbers were records, but this isn't necessarily statistically meaningful. Some scientists conclude that such increases come from global warming, but others say they represent a natural cycle, so the jury is still out. What is clearer is that even if global warming isn't increasing the number of hurricanes, it is probably increasing their intensity, as the 2007 IPCC report states. Hurricanes form over oceans, and the warmer the water, the more energy and destructive potential the storm carries. In any case, *The Day After Tomorrow* stacks the deck by showing every kind of intense phenomenon, including a huge tsunami, arising in a short time. One effect it shows is impossible, hurricane-like storms with central eyes forming over land.

The freezing of the Northern Hemisphere in the movie is, however, based on the way the world really works. The ocean near the equator absorbs heat from the sun, and the Gulf Stream carries the heated water into the North Atlantic, keeping the northeastern United States and Europe warm. As the water gives up its heat, it becomes denser and sinks, turning into a cold river under the ocean that later flows back to the south, where it resurfaces to be warmed again. This watery conveyer belt is known as the Thermohaline Circulation, or THC.

Since the density of the water matters, fresh water, which is less dense than salt water, can disrupt the THC, as is thought to have happened in the past.

Some researchers believe that 8,000 years ago, melting ice in Canada produced a huge body of fresh water that poured into the North Atlantic. The influx shut down the THC and produced an ice age that lasted two centuries in Northern Europe. This is the scenario foreseen by Jack Hall, except that in the film, the fresh water comes from increased rain and the melting of ice because of global warming.

There are two problems with how this is presented in *The Day After Tomorrow*. Even when the film was made, many climate scientists thought that a major change in the THC was unlikely during this century. Now newer evidence shows that climate change is indeed affecting the THC, but not enough to cause an ice age. The 2007 IPCC report projects that the circulation of warm water northwards will decrease 25 percent by 2100. The projections also show, however, that rising atmospheric temperatures will more than offset the cooling effect, so an ice age is considered out of the question in this century. The other issue is the speed of the THC breakdown, which would occur over decades and not days, should it ever happen. But there's one qualification. In a system as complex as the Earth, with many interacting elements, there's some small chance it'll go nonlinear; that is, past a certain threshold, the system's responses become disproportionate to what's causing them. When nonlinearity sets in, then all bets are off as to the size and speed of what happens next.

Like films about aliens and cosmic collisions, films about our planet's natural changes reflect different levels of speculation. Although each film has its scientific seed, how each lets the seed develop defines different approaches. Some, like *Twister* and *Dante's Peak*, stay within accepted science or extend it only slightly, not entering unknown or impossible territory. Some distort the science to create drama, good visuals, or satisfying scenes of destruction; *Volcano* does this with basic geology, so we can watch the spectacle of a volcanic eruption in Los Angeles. Some achieve the same cinematic ends not exactly by distorting the science but by picking the most dramatic scenarios, even if they are highly improbable: *The Day After Tomorrow* does this so its events can unfold at a breathtaking clip. And some, like *The Core*, quickly leave the scientific seed behind and make up the rest of the science as they go along.

The different levels of speculation in these films arise partly because we don't fully grasp the complex natural processes of our planet, from the formation of tornadoes to climate variation. The next chapters present disasters of our own making: nuclear accidents, unfortunate tinkering with genes, and computers that go out of control. As our own creations, arising out of our own science and technology, we might hope to understand these events better and express them directly in the movies. But do we really fully grasp and control our human-made world?

PART II

DANGERS FROM OURSELVES

Chapter 5

Atoms Unleashed

Julian Osborne: Who would ever have believed that human beings would be stupid enough to blow themselves off the face of the Earth?

—*On the Beach* (1959)

Greg Minor [after the Ventana nuclear power plant almost undergoes meltdown]: You're probably lucky to be alive . . . I think we might say the same for the rest of southern California.

—*The China Syndrome* (1979)

Jack Ryan [as the United States and Russia are about to launch nuclear strikes]: General, . . . if you shut me out, your family, and my family, and twenty-five million other families will be dead in thirty minutes.

—*The Sum of All Fears* (2002)

Wrenching climatic change, collision with an asteroid, a meeting with aliens—any of these events could change the world. But apart from such possibilities, we've already changed our world through the discovery of nuclear fission, the splitting of the atom. This huge achievement, a major scientific breakthrough of the twentieth century, enlarged our understanding of the universe and affected how we live. It supported a military technology of mass destruction that altered the course of World War II, dominated Cold War politics for decades, and raised deep ethical questions. Now, with the Cold War over, there's still great concern about nuclear weapons and nuclear materials in the hands of rogue nations or terrorists, as shown by the weight given to the possibility that Saddam Hussein had a nuclear capability in Iraq.

For obvious reasons, films have followed the power of the atom almost from its beginnings. Nuclear fission was first observed in 1938, by Otto Hahn and Fritz Strassman at the Kaiser Wilhelm Institute for Chemistry in Berlin. They bombarded uranium with neutrons, but it wasn't obvious that the uranium nucleus had truly split until further analysis by Lise Meitner and Otto Frisch. Soon after, scientists recognized the potential to use nuclear fission in a powerful weapon. Since the fission breakthrough occurred under the Nazis, fears that they would develop an atomic bomb spurred American efforts to build the weapon first. (In an ironic twist, Meitner had worked with Hahn for years in Berlin but had to leave Germany in 1938 because of her Jewish heritage.)

From the cinematic viewpoint, few other real-life scientific stories can match the drama, tension, and high stakes of the Manhattan Project to build an atomic bomb during World War II. On the speculative side, the potential for nuclear weapons, nuclear accidents, and nuclear terrorism to cause widespread devastation opens imaginative though mostly pessimistic story possibilities.

The atomic bomb was only five years old when the modern era of science fiction film began in 1950. Created and tested in New Mexico, at Los Alamos and the Trinity site at Alamogordo, then dropped on Hiroshima and Nagasaki in 1945, the bomb's horrific effects were still being absorbed. Some films centered on its blast and heat that caused widespread death and destruction; others featured the frightening, almost magically evil effects of nuclear radiation, invisible and lethal. Still others pointed out the moral dilemmas surrounding this destructive weaponry. These issues became more pressing as we entered deeper into the nuclear age with even more destructive thermonuclear weapons, that is, hydrogen bombs. These use the process of fusion, in which hydrogen nuclei unite to form helium with a massive release of energy. As the United States and Soviet Union squared off against each other with nuclear-tipped intercontinental missiles, movie scenarios portrayed vast destruction, including the death of all humanity.

But atomic energy from fission wasn't only used for war. It entered daily life, not in vehicles and aircraft as predicted by some futurists (and a good thing, too) but in nuclear reactors that proved practical power sources, until accidents at Three Mile Island and especially at Chernobyl highlighted the dark side of these installations, which was then reflected in films. Meanwhile, as the Cold War wound down and threats of nuclear exchanges on the grand scale decreased, other films focused on nuclear destruction inflicted by terrorists. Still other movies reflected the latest nuclear technology, the effort to produce power through nuclear fusion. While this is the process that drives a hydrogen bomb, it also powers the stars and our own Sun and could, it's hoped, provide clean and limitless energy on Earth as well.

The Manhattan Project became open knowledge after World War II ended, and *The Beginning or the End*, about the project, came soon after, in 1947. It featured fictionalized versions of J. Robert Oppenheimer (Hume Cronyn), the scientific director of the Los Alamos laboratory; Maj. Gen. Leslie Groves (Brian Donlevy), the army engineer in command of the Manhattan Project; and other leading real-life characters. Other films (and television treatments) followed, enough of them to trace out a film chronology of the real and then the imaginary history of nuclear power and weaponry. Two movies based on real history are the documentary *The Day After Trinity* (1981) and the drama *Fat Man and Little Boy* (1989).

The Day After Trinity uses contemporary interviews with participants in the Manhattan Project and archival footage of Oppenheimer (he died in 1967), Los Alamos, the Trinity test, and Hiroshima after the blast. It tells the story of Oppenheimer, the brilliant scientist-scholar-administrator who made the Manhattan Project work—and whose security clearance the government revoked in 1954 because of communist associations and because he was ambivalent about building a hydrogen bomb and espoused international control of atomic weaponry. The film's title comes from a 1960s interview in which Oppenheimer, asked about international control, replied "It's twenty years too late. It should have been done the day after Trinity."

Fat Man and Little Boy is a fictionalized story of the Manhattan Project; its title comes from the code names for the atomic bombs dropped on Japan. It portrays Oppenheimer (Dwight Schultz), Groves (Paul Newman), and other major figures, such as Leó Szilárd (Gerald Hiken), the Hungarian-born scientist who helped set the stage for the Manhattan Project but later turned against the use of the bomb. The film covers much the same ground as *The Day After Trinity*—the building of the bomb and its political ramifications, the thinking behind the use of the bomb against Japan, and the misgivings that Oppenheimer and other physicists came to feel.

Unlike *The Day After Trinity*, the movie also presents something about the scientific and technical problems that arose in constructing the weapon and shows the flashes of brilliance that solved them. It adds real human elements, such as the relationships between Oppenheimer and his wife and mistress, and fictional ones, such as a love story between a nurse (Laura Dern) and imaginary Los Alamos scientist Michael Merriman (John Cusack), who dies a horrific death by radiation.

These semirealistic portrayals of nuclear weapons and scientists have been accompanied by a parallel thread of fictional movies, some in classic science fiction form, others projecting nuclear war in contemporary times or the near future. All closely reflect the attitudes of the time. Fear of nuclear destruction permeated society beginning in the 1950s and throughout the Cold War, an era when school children were drilled against nuclear attack in "duck and cover" exercises and people built backyard fallout shelters. Similarly, the power of the atom permeated many films, even those about other subjects.

For example, two classics about alien encounters, the original *The Thing from Another World* and *The War of the Worlds*, show Geiger counters being used to check radiation from the unwelcome visitors, though there's no particular reason to think they're radioactive. In *Colossus: The Forbin Project* (1970) and the Terminator series, the computers that aspire to control or erase humanity are initially given control over weapons to defend the United States against nuclear attack.

Other films use nuclear effects directly. In *Godzilla*, nuclear blasts awaken the monster of that name, who had been asleep near Bikini Atoll, in reality the site of much U.S. nuclear testing. In *Them!* (1954), radiation from the Trinity atomic bomb test mutates ordinary ants into gigantic forms that threaten to take over the world. In the original 1953 version of *The War of the Worlds*, the military uses its ultimate weapon, an atomic bomb, against the invaders (although the Martian base is near Los Angeles, no one seems worried about collateral damage).

More recent films also use nuclear devices to solve all sorts of problems, but not always effectively. In *The Core*, scientists explode hydrogen bombs to set the Earth's core spinning again, although this undirected violence couldn't possibly nudge the core into rotation. In *Meteor*, *Deep Impact*, and *Armageddon*, the knee-jerk reaction is to nuke the incoming comet or asteroid with a hydrogen bomb or two, with no guarantee that this will resolve the crisis.

Although some films use the sheer unthinking violence of a nuclear bomb, *The Day the Earth Stood Still* (1951) is different, presenting moral issues instead of explosions. The story begins as a saucer-shaped craft approaches Earth from space and lands on the Mall in Washington, D.C. The government rings the ship with troops, and everyone waits breathlessly as a hatch opens. What emerges is not some repulsive alien, but a human-appearing being in a silvery jumpsuit. The visitor, Klaatu (Michael Rennie), says in excellent English "We have come to visit you in peace," but when he extends his hand with something in it, a trigger-happy soldier shoots and wounds him. In response, a tall, hulking humanoid robot (played by Lock Martin) appears from the spaceship and emits a ray that destroys rifles, tanks, and artillery at a touch. The object in his hand, says Klaatu after he's shot, was a gift for the U.S. president.

Klaatu quickly recovers, and tells the president's representative, Mr. Harley (Frank Conroy), that he has an urgent message for the leaders of all the Earth's nations. Harley insists that the international climate makes such a meeting impossible, and Klaatu, frustrated by the violence, squabbling, and stupidity he's encountering, escapes from the hospital. While headlines blare "Man from Mars Escapes," Klaatu rents a room under the name Carpenter and makes friends with widowed Helen Benson (Patricia Neal) and her young son Bobby (Billy Gray).

Since the Earth's governments won't listen, Klaatu gives his message to Prof. Jacob Barnhardt (Sam Jaffe), the world's leading scientist: if the Earth continues on a course that could threaten other planets by taking atomic weapons into space, it will be eliminated. Barnhardt agrees to gather other scientists to hear Klaatu, but first, to show the power he commands, Klaatu makes the Earth "stand still" by turning off electricity around the world for a time, sparing only critical applications like hospitals.

Meanwhile, the government is closing in, and Klaatu confides in Helen, who quickly grasps the importance of his mission. If anything happens to him, he says, she must go to the robot Gort and say "Klaatu barada nikto." When a military patrol fatally shoots Klaatu, Helen makes her way to the spaceship, fearfully approaches the robot, and says the words. Gort responds by retrieving Klaatu's body and placing it in a reanimating apparatus aboard the ship. As Helen watches in awe, Klaatu begins breathing and recovers, though he tells Helen the return to life is only temporary. Then he gives the assembled scientists his message, explaining that his home world is part of an organization of planets and that

> we created a race of robots [to] preserve the peace. . . . At the first sign of violence they act automatically against the aggressor. . . if you threaten to extend your violence this Earth of yours will be reduced to a burned-out cinder. . . . Join us and live in peace [or] pursue your present course and face obliteration. . . . The decision rests with you.

After these words, which leave the crowd impressed and thoughtful, Klaatu gives Helen a half wave, she returns a nervous smile, then he and Gort vanish into the spaceship. It starts glowing, and we hear a hum of power as the crowd scatters; then the ship lifts straight up and vanishes rapidly into space.

The Day the Earth Stood Still makes its point without showing nuclear destruction; other early films extrapolated how a nuclear exchange might begin and where it might lead, as in *Dr. Strangelove, or: How I Learned to Stop Worrying and Love the Bomb* (1964), *Fail-Safe* (1964), and *On the Beach* (1959). *Dr. Strangelove* paints an absurdist picture of nuclear war. Starting with scenes of huge U.S. strategic bombers, we soon see that something is badly awry. General Jack D. Ripper (Sterling Hayden), commander of Burpelson Air Force Base, has gone mad and sent an aircraft wing to attack the Soviet Union. He's been obsessed by water purity since fluoride started being added to water to prevent tooth decay, and says, "I can no longer allow . . . the international Communist conspiracy to sap and impurify all of our precious bodily fluids."

Aboard one of the bombers, its commander Major T. J. "King" Kong (Slim Pickens) questions the attack order but finally accepts it, puts on a cowboy hat, and tells his crew to prepare for "nuclear combat toe-to-toe with the Russkies." Meanwhile General Buck Turgidson (George C. Scott) hurries to the War Room to consult with President Merkin Muffley (Peter Sellers) and his advisers. Among them is wheelchair-bound, former Nazi scientist Dr. Strangelove, in charge of U.S. weapons research and development, whose bionic arm occasionally snaps up into the Nazi salute (Strangelove is played by Sellers as well, who also plays General Ripper's British executive officer, Group Captain Lionel Mandrake).

The president calls Soviet Premier Kissof to say that "one of our base commanders . . . went a little funny in the head. . . . And, ah . . . he went and did a silly thing," then learns that the Soviets have built a computer-controlled

Dr. Strangelove, or: How I Learned to Stop Worrying and Love the Bomb (1964). In this dark Cold War comedy about nuclear annihilation, all attempts fail to recall a U.S. strategic bomber sent by a crazed Air Force general to attack the Soviet Union. The aircraft commander, Major T. J. "King" Kong (Slim Pickens, in the cowboy hat) zestfully evades Soviet air defenses and drops a hydrogen bomb that triggers a worldwide nuclear doomsday.

Source: Hawk Films Prod/Columbia/The Kobal Collection.

nuclear doomsday device that will automatically wipe out all life if they are attacked. Meanwhile, Mandrake guesses the recall code for the stray bombers and they return, except for Kong's. Reaching his target, Kong straddles a hydrogen bomb and rides it down like a cowboy, whooping and waving his hat, only to dissolve in a mushroom cloud of destruction. In the next scene, Dr. Strangelove proposes that government and military leaders can survive doomsday in mine shafts, with ten women to each man for breeding purposes. The film ends as Strangelove rises from his wheelchair, totters toward the president, and says "Mein Führer! I can walk!" followed by scene after scene of immense nuclear mushroom clouds.

The movie *Fail-Safe* (1964), from a novel by Eugene Burdick and Harvey Wheeler, presents a similar scenario, but without any dark humor. Technical glitches mistakenly send a squadron of nuclear-carrying Vindicator bombers to attack Moscow. They can't be recalled, and although the U.S. president (Henry Fonda) tells the Soviets how to shoot down the aircraft, one skillfully flown bomber evades Soviet air defenses. As it bores in toward Moscow, some of the president's inner circle urge him to exploit the situation and finish off the enemy. Rejecting this, the president offers the Soviet premier a desperate and brutal eye-for-an-eye solution that the premier accepts as the only way to avoid global war: the president orders a U.S. bomber to take out New York City in compensation for the loss of Moscow.

On the Beach (1959) continues in a serious vein, showing the tragic aftermath of a nuclear war such as might have been started by a *Fail-Safe* error or a General Ripper. Based on a novel by Nevil Shute, the movie is set five years in the future. The U.S. nuclear submarine *Sawfish*, under Captain Dwight Towers (Gregory Peck), docks at Melbourne, Australia. The *Sawfish* has survived a nuclear war that began for reasons unknown and has devastated most of the world. Australia escaped direct damage, but scientists predict that a cloud of deadly radioactive dust will arrive in five months, wiping out this last enclave of humanity.

The prospect is so grim that suicide pills are supplied to those who want to avoid a miserable death by radiation. However, one scientist offers hope, suggesting that rain and snow have washed out the radioactivity. To test this, Tower takes the *Sawfish* north, along with physicist Julian Osborne (Fred Astaire). They will also investigate a puzzling radio signal in Morse code that comes from San Diego, which seems to indicate someone is alive there.

Off the Alaskan coast, Osborne finds that the radiation level remains lethal. Though this is a bad blow, Tower and his crew turn toward San Diego, examining an intact but eerie, devastatingly empty San Francisco along the way. Continuing south, one of the naval officers asks Osborne, the resident "egghead," "Who do you think started the war?" "Albert Einstein!" answers Osborne, adding: "Everyone had an atomic bomb and counter bombs and counter counter bombs. I know, I helped build them. . . . [Somebody] probably looked in a radar screen and thought he saw something . . . he pushed the button and . . . and . . . the world went crazy." Arriving in San Diego harbor, a crew member goes ashore and finds the source of the mysterious radio signal: not a person, but a fluke. A soft-drink bottle has become entangled in an

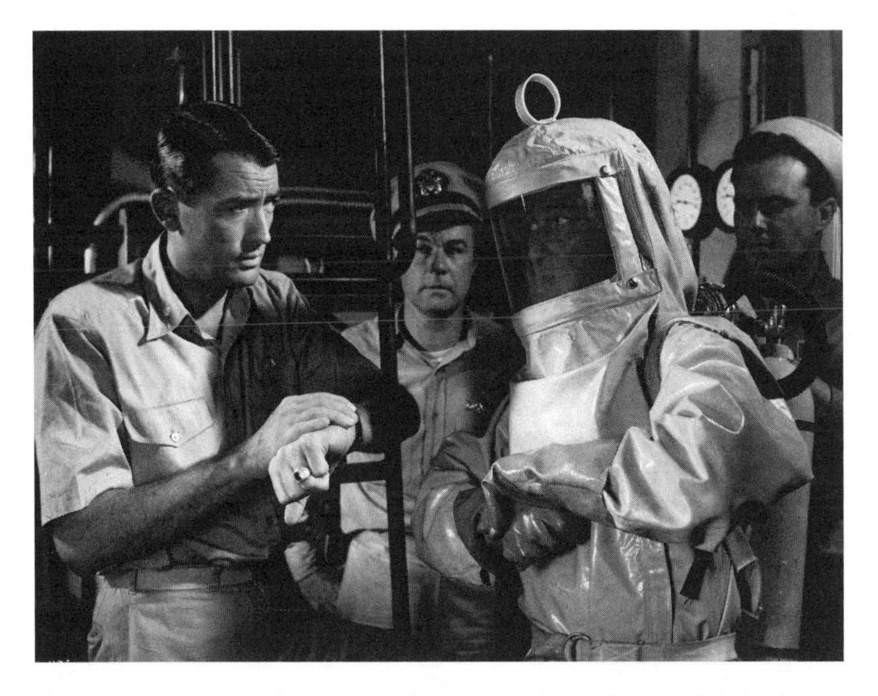

On the Beach (1959). Nuclear submarine captain Dwight Towers (Gregory Peck, left) instructs radiation-suited Lt. Sunderstrom (Harp McGuire) before Sunderstrom goes ashore to seek survivors after a worldwide nuclear exchange. The search is unsuccessful, making this bleak film about nuclear holocaust in the near future even grimmer. *A Golden Eagle; see chapter 9.*

Source: United Artists/The Kobal Collection.

old-fashioned window shade and, leaning against a radiotelegraph key, makes random "dits" and "dahs" as the wind blows.

With all hope gone, there's nothing to do but return to Melbourne and face the end along with everyone else. Osborne fulfills a life-long dream, driving a Ferrari to win a brutal race in which one reckless driver after another, knowing the end is coming anyway, dies in a fiery crash. As radiation sickness begins to appear, some turn to religion, and families line up for the poison pills, but Osborne dies smiling in his Ferrari, revving it up to fill the garage with fumes.

Tower sets aside memories of his dead family and shares some fleeting happiness in a love affair with beautiful Moira Davidson (Ava Gardner), but he's also pulled by duty. He accepts his crew's request to sail back to the United States and die there, although it means leaving a fearful Moira all alone. She watches from a headland as Towers, aboard the *Sawfish,* looks up at the sky before the submarine dives on its journey home. The final images show Melbourne utterly silent and deserted, with papers blowing forlornly in the streets, an abandoned trolley car, and a sign remaining from the religious gatherings: "There is still time . . . brother."

A television remake of *On the Beach* (2000) substantially followed the earlier version, with some changes. In this story, scientists argue vehemently about whether the radiation has diminished, raising hopes that some 1,000 people can be evacuated north. But when the U.S. nuclear submarine *Charleston* heads to Alaska, its crew finds no reduction in radiation. Also like the original, a mysterious message is a fluke, although in this up-to-date version, it's due to sunlight turning on a solar-powered computer, not a randomly tapping radiotelegraph key. The 2000 version is more graphic in showing San Francisco destroyed, in displaying radiation-induced vomiting and skin sores, and in showing violent breakdowns of order in Melbourne. But unlike the original, when the submarine crew elects to put out to sea and die, their captain (Armand Assante) returns to his lover Moira (Rachel Ward) to be with her at the end.

Other science fiction films painted a slightly less awful outcome, imagining a postnuclear world left reeling, with millions or billions killed and civilization destroyed, but with some survivors. *Five* (1951) takes us almost to the Adam and Eve minimum that can repopulate the world, with only four men and a pregnant woman, from bank teller to neo-Nazi, surviving a nuclear holocaust. *A Boy and His Dog* (1975), from a story by Harlan Ellison, begins with nuclear blasts and then informs us that "World War IV Lasted Five Days." Now it's 2024, and the war has left the southwestern United States a barren wasteland. Among the gangs roaming this desert, lone drifter Vic (Don Johnson) and his telepathic dog, Blood (voice of Tim McIntire)—presumably a radiation-induced mutation—search for food and women; encounter Screamers, who seem to be mutants that glow green; and find an underground society of survivors that looks normal but isn't. The Mad Max series (*Mad Max* [1979], *Mad Max 2* or *The Road Warrior* [1981], and *Mad Max Beyond Thunderdome* [1985]) takes place in Australia after a nuclear holocaust. Small enclaves of people live in the vast empty spaces of the Outback, maintaining a few remaining bits of civilized life and technology; in *Road Warrior,* "Mad" Max Rockatansky (Mel

Gibson) helps a group that operates a gasoline refinery save their precious fuel from a bizarre gang of marauding bikers.

As the Cold War wound down, nuclear end-of-the world stories became less prevalent in the movies, which could even treat nuclear weaponry with a dash of comedy. In *The Manhattan Project* (1986), teenage genius Paul Stephens (Christopher Collet), a star in his high school science class, decides he can build his own atomic bomb as the world's most impressive science fair entry. With the help of his girlfriend Jenny (Cynthia Nixon), he does just that, but matters turn serious when weapons physicist John Mathewson (John Lithgow) and the U.S. government become involved.

The collapse of the Soviet Union in 1991 further reduced the chances of superpower nuclear interchanges that would desolate the whole Earth, but nuclear dangers from accidents and worse still remained. The events of 9/11 showed how terrorists can subvert technology to cause enormous death and destruction. Even with the Cold War over, the five biggest nuclear-power states (the United States, Russia, the United Kingdom, France, and China) together stockpile thousands of nuclear weapons. Israel, North Korea, and Iran are thought to possess nuclear weapons or to have the ability to make them. The nuclear nations also maintain huge stocks of weapons-grade plutonium and uranium. Then there are civilian nuclear reactors, whose nuclear fuel could be used to make dirty bombs, in which a conventional explosive like dynamite is used to widely spread toxic or lethal radioactive material. To say the least, not all these weapons and materials are carefully guarded or accounted for. Whether terrorists use spent reactor fuel or steal or make nuclear weapons, nuclear terrorism is a worrisome possibility.

Beginning with the accidental death by radiation of a scientist in *Fat Man and Little Boy*, movies have projected the consequences of accidents, mismanagement, and terrorism. *The China Syndrome* is an early film about nuclear mishap that appeared in 1979, only days before the real Three Mile Island accident. Kimberly Wells (Jane Fonda) is a popular Los Angeles TV feature reporter who yearns to cover hard news. For a series about energy sources, she and her long-haired cameraman Richard Adams (Michael Douglas) visit the Ventana nuclear power plant. Company PR man Bill Gibson (James Hampton) gives them a reassuring spiel about nuclear power and how the huge concrete dome called the containment is designed to safely house 20 million pellets of uranium fuel in the reactor.

The China Syndrome (1979). Shift supervisor and nuclear submarine veteran Jack Godell (Jack Lemmon) discusses serious problems at the Ventana nuclear power plant with TV reporter Kimberly Wells (Jane Fonda). Soon after its release, the film's message about the dangers of nuclear power was unexpectedly underscored by the real Three Mile Island nuclear accident in Pennsylvania.

Source: Columbia/The Kobal Collection.

But as they tour the plant, Kimberly and Richard feel a sudden vibration, hear emergency horns, and see panic among control room personnel who are watching an apparently rising but really rapidly decreasing level of cooling water for the reactor. The problem, partly attributable to a faulty readout, is brought under control by shift supervisor Jack Godell (Jack Lemmon), a veteran of nuclear submarines. Richard, who mistrusts nuclear power, surreptitiously films the scene.

The power company minimizes the event, but independent scientists viewing Richard's film tell Kimberly that Ventana came close to complete disaster.

If the falling water level had uncovered the reactor, its incredibly hot core would have melted through the earth to China. But first, when it reached groundwater, it would have produced enough atmospheric radioactivity to desolate much of southern California. Meanwhile, Jack investigates further and finds that the welds on the water pumps are substandard; the X rays meant to confirm that they're safe have been falsified.

The company is desperate to conceal this damaging information, and its security men attack a messenger carrying the falsified X rays and take them. With the evidence gone, Jack takes over the Ventana control room at gunpoint, and Kimberly puts him on TV to say the plant is unsafe. The company orders a SCRAM—an emergency reactor shutdown—to distract Jack, but a SCRAM is just what Jack feared would stress the bad welds. Amid blaring emergency horns, a SWAT team shoots Jack, while the welds break as he predicted. A main pump teeters on the edge of catastrophe; the system barely manages to stabilize, although Jack dies.

Later, PR man Gibson announces that the public was never in any danger and portrays Jack as emotionally disturbed and drunk. But Kimberly, on live TV, asks Jack's second in command, Ted Spindler (Wilford Brimley), if he thinks the plant should be shut down. Gibson tries to shut Ted up, but he replies: "Jack Godell was . . . was my best friend. I mean these guys are painting him as some kind of a loony . . . . He was the sanest man I ever knew . . . he wouldn't have done what he did if there weren't something to it. . . . Jack Godell was a hero." Kimberly adds on air, "what happened tonight was not the act of a drunk or a crazy man . . . let's just hope it doesn't end here," as the film itself ends.

Silkwood (1983) also raises issues of safety violations and cover-ups in the nuclear industry. Based on a true story, it takes place at a plant in Oklahoma that makes fuel rods for reactors. Karen Silkwood (Meryl Streep) is one of the workers who deals with the mixture of plutonium and uranium compounds that go into the rods. Although radiation levels are monitored, the company is running late on its contracted delivery dates, and Karen sees evidence that the pressure to deliver is causing safety issues. She also learns that the company hasn't been straightforward about the risks of working with plutonium and that the standards for "safe" levels of contamination are dubious: as one doctor says, there is no safe level.

Worst of all, as in *The China Syndrome*, she finds that X-ray negatives used to check welds in the fuel rods are being falsified. These fuel rods are made for a breeder reactor—that is, one that produces more nuclear fuel—in Hanford, Washington. As Karen hears from union leader Max Richter (Josef Sommer):

> In an ordinary nuclear plant you can have meltdowns, poisonous gas, and dead people. That's nothing compared to what can go wrong with a breeder. You put defective fuel rods into a breeder reactor . . . for all we know, the whole state could be wiped out . . . they could kill off two million people. There's a moral imperative involved here.

Karen gathers information about safety violations and the X-rays, but before she can talk to a *New York Times* reporter, she's found to be contaminated with plutonium herself, under mysterious circumstances; and then is found dead after her car runs off the road. That was the fate of the real Karen Silkwood, whose death, though ruled a single car accident, remains an unsolved mystery to some.

Beyond accident and cover-up, there's nuclear terrorism, which could deliver either a nuclear bomb or a dirty bomb. In the movie *Critical Mass* (released on DVD in 2000), terrorists with ties to right-wing militia groups in the United States employ both approaches. They steal the makings of an atomic bomb, which they plan to assemble and explode inside a decommissioned nuclear power plant to spread additional fissionable material around the countryside; their plot, however, is foiled.

In *The Sum of All Fears*, a working atomic bomb is the instrument of terror. The weapon is buried in the desert, intact, when the Israeli aircraft carrying it is brought down by a Syrian missile during the 1973 Yom Kippur war. Long after, the bomb is retrieved by an arms dealer who sells it to a group of neo-Nazis. Their leader, Richard Dressler (Alan Bates), has a plan: instead of stupidly fighting both America and Russia, as Hitler did, Dressler will set the two superpowers at each other's throats by exploding the bomb in the U.S.; then the fascists can step in as a world power.

The neo-Nazis set off the bomb at the Super Bowl with much loss of life, and as planned, U.S. president Robert Fowler (James Cromwell) assumes that Russian president Nemerov (Ciarán Hinds) is responsible. Fowler prepares to launch a major nuclear reprisal against Russia, but fortunately CIA analyst

Jack Ryan (Ben Affleck) steps in. Jack's been taken under the wing of CIA director William Cabot (Morgan Freeman) because Jack has made a special study of Nemerov and is convinced that he would not launch such an attack. With the help of his CIA colleagues, Jack shows the president that the Russians aren't responsible. Apocalyptic nuclear strikes and counterstrikes are aborted at the very last instant, and as the film ends, the United States and Russia are signing a treaty pledging to eliminate rogue weapons globally, while the neo-Nazis are hunted down and killed.

The nuclear fission in these movies, which splits heavy atomic nuclei into lighter ones, is a well established technology. That's not the case for nuclear fusion, which produces energy by merging light atomic nuclei into heavier ones. Although we use fusion in the hydrogen bomb, despite decades of effort scientists have yet to achieve a practical civilian version by either of two methods: hot fusion, which requires temperatures of millions of degrees or immense lasers; and cold fusion, which for a brief time seemed possible at ordinary room temperatures. These failures haven't kept screenwriters from exploiting the idea. In *Spider-Man 2* (2004), physicist Dr. Otto Octavius (Alfred Molina) works on a hot fusion experiment that goes badly wrong and turns him into a cyborg, as I describe in chapter 7. Fusion also appears in *Chain Reaction* (1996), although in garbled form, and in *The Saint* (1997).

Chain Reaction opens with aerial shots showing massive atmospheric contamination spewing from huge smokestacks. Then Dr. Alistair Barkley (Nicholas Rudall) tells a large audience that the world needs a pollution revolution, a clean energy technology. Barkley runs the Hydrogen Energy Project, which is funded by a foundation under Paul Shannon (Morgan Freeman) and employs technician Eddie Kasalivich (Keanu Reeves) and physicist Dr. Lily Sinclair (Rachel Weisz). The project fails to extract power from the hydrogen in water until Eddie accidentally discovers a certain set of sound frequencies that make the reaction chamber bubble furiously and start producing a stable flow of energy.

Eddie later returns to the laboratory and finds that Barkley has been murdered and that the hydrogen power is approaching overload. Racing on his motorcycle, he barely outpaces a massive explosion that levels eight city blocks, set off by a mysterious group that had earlier planted a bomb. The FBI finds evidence implicating Lily and Eddie, who realize they're being framed and go into hiding. Meanwhile, the FBI investigates Shannon's connection

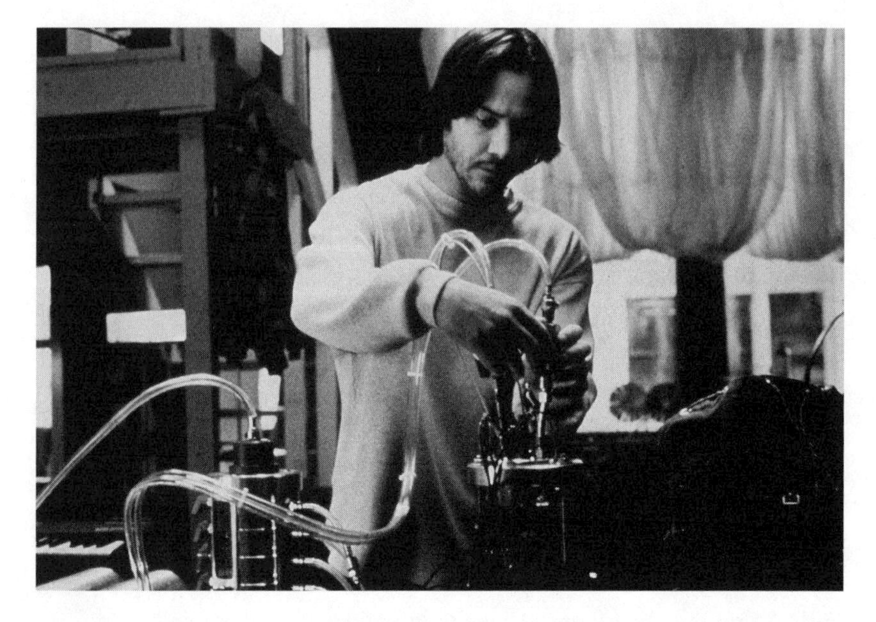

Chain Reaction (1996). Laboratory technician Eddie Kasalivich (Keanu Reeves) outdoes the physicists and finds a way to produce power from nuclear fusion. Although the film confuses different approaches to fusion, it represents a shift, from the nuclear terrors of earlier films like *On the Beach* to a benign view of unlimited, nonpolluting nuclear power for humanity.

Source: 20th Century Fox/The Kobal Collection/Deana Newcomb.

with C-Systems Research, a shadowy operation with governmental connections and an enormous underground complex in Virginia.

Eddie finally figures out that Shannon is playing a double role, but he is too late to prevent C-Systems agents from kidnapping Lily to try another hydrogen energy experiment, which, as Eddie watches, fails. He reconfigures the experiment with the right frequencies to produce energy at an escalating rate. Then he confronts Shannon over Barkley's murder, which Shannon justifies by saying that hydrogen technology would cause huge economic disruption. Finally, the hydrogen reaction explodes and destroys C-Systems, although Eddie escapes with Lily. FBI agents soon arrive to tell the two they're in the clear, while Eddie explains that he's sent the plans for clean, cheap energy to thousands of scientists. Shannon also escapes and drives away in a limousine, his role still ambiguous, while Eddie and Lily fly off in an FBI helicopter.

Fusion, the cold variety, is also central in *The Saint.* Ivan Tretiak (Rade Serbedzija), a Russian oil and gas billionaire, has ambitions to head a new Russian empire. He's created a shortage of heating oil, and by using the limitless power of cold fusion to relieve the suffering of millions of freezing Russians, he'll win popular support for a coup against the Russian president.

Although cold fusion hasn't worked yet, American electrochemist Dr. Emma Russell (Elisabeth Shue) has passionate faith in the process. In a speech to Oxford University science students, she glows about its promise. "Just imagine," she says, "there's more energy in one cubic mile of seawater than in all the known oil reserves on Earth . . . you could drive your car 55 million miles on a gallon of heavy water . . . it would be the end of pollution, warmth for the whole world."

Emma has made a breakthrough that will turn cold fusion into practical reality, and Treviak hires Simon Templar (Val Kilmer), the Saint, to steal her process. Simon is a millionaire thief who employs ingenious disguises and high-tech devices in his crimes. He learns that Emma is a romantic, a lover of the poetry of Shelley and Byron, and is searching for the right man in her life. Simon seduces her, steals her notes, and passes them to Treviak. But despite his emotional detachment, he falls in love with her.

After obtaining the secret, Treviak tries to kill Simon and Emma while planning his coup against the Russian government. Simon foils this when he gains entry to the Russian president and arranges matters so that Treviak's use of cold fusion backfires. Treviak is arrested, Russian democracy is saved, and Simon and Emma go off together. Simon, however, wants to make money from Emma's discovery, but she convinces him that giving it away would be the right thing. As Emma makes a speech presenting cold fusion to the world, the Saint eludes the police one last time. As he drives away, we hear on the car radio that an anonymous donor—Simon, of course—has funded the nonprofit Russell Foundation, devoted to providing free energy to all.

Fusion power flashed strongly on the public consciousness, and nuclear fission has been embedded there since the advent of the atomic bomb, which quickly taught citizens the rudiments of the science. Most people know something about Los Alamos, megatons of explosive power, uranium and plutonium, and critical mass; everyone knows radiation has bad effects. Against this background, how accurately do the films follow the science?

The documentary or semi-documentary films do reasonably well. They get basic facts about fission right and sometimes dig deeper. *Fat Man and Little Boy* identifies a real technical problem that arose at Los Alamos, how to smash together or compress chunks of fissionable material fast enough to make an explosion. It shows in some detail the two approaches considered, the gun method and the implosion method, and how they worked out. On the civilian side, *The China Syndrome* presents a minilecture on the workings of a nuclear power plant, which is enough for the audience to follow the evolution of the crisis.

In portraying the terrible death by radiation of fictional scientist Michael Merriman, *Fat Man and Little Boy* also reflects radiation accidents that killed two real Los Alamos physicists, Louis Slotin and Harry Daghlian. The film slightly muddles the causes of the accidents and dramatically intercuts Merriman's illness and death with scenes from the Trinity test in July 1945, whereas Daghlian and Slotin died later. But the movie shows how such accidents can happen and the symptoms of acute radiation sickness. These symptoms develop within minutes to days after exposure to radiation and include vomiting, diarrhea, and skin damage, all of which are portrayed on-screen for Merriman. Other injuries may include a reduced blood cell count, male sterility, and a clouding of the lens of the eye, but if enough radiation is absorbed, death results from damage to the bone marrow, nervous system, or gastrointestinal system. In the movie, Merriman's gastrointestinal system is said to be destroyed, consistent with the high dosage he received.

Films have also been reasonably accurate and sometimes prescient in projecting radiation accidents that affect large numbers of people, as in the Three Mile Island and Chernobyl events. The Three Mile Island nuclear plant near Middletown, Pennsylvania, was the site of the worst nuclear disaster in the United States. On March 28, 1979, the reactor developed serious problems eerily like those described in *The China Syndrome*, released barely two weeks earlier, and that had coincidentally likened the scope of possible disaster in southern California to the area of Pennsylvania. Failure of the main pumps at Three Mile Island led to a lack of cooling water, although instrumental readouts didn't show this. The nuclear fuel overheated; half the reactor core melted; but the containment vessel held, avoiding a catastrophic deluge of radiation. Even so, some radioactive gas was emitted, to the panic and confusion of the surrounding population over a period of several days, but no lives were lost or injuries sustained.

That wasn't true for the world's worst nuclear power accident, at the Chernobyl plant in what was then the Soviet Union and is now Ukraine. On April 26, 1986, during a system test, operators ignored safety requirements, and the reactor went out of control. The power output jumped to ten times normal, producing an explosion that released a massive dose of radiation and a large fraction of the nearly 200 tons of radioactive material in the reactor. The official count for the immediate death toll was thirty-one, mostly firefighters and rescue workers exposed to the radiation. Nearly 150,000 people within an eighteen-mile radius had to be evacuated, and the radioactive plume was detected as far away as the United Kingdom and the United States. The long-term effects on the inhabitants and the environment of the region are still subject to debate, although there is evidence of an increase of thyroid cancer in children.

Films that project even greater disasters from all-out nuclear war, such as *On the Beach*, also reflect real possibilities. In his 1961 book *On Thermonuclear War*, military strategist Herman Kahn considered the outcome if the United States and the Soviet Union went to war with hydrogen bombs. Applying supposedly objective analysis, and introducing the term "megadeath," or one million deaths due to nuclear weapons, he compared strategic scenarios with "only" ten megadeaths to those with one hundred megadeaths. His approach dismayed and repelled many, but similar thinking appears in *Dr. Strangelove*. General Turgidson has a binder labeled "World Targets in Megadeaths," and at one point, urges the president to follow up on General Ripper's attack and catch the Soviets "with their pants down." In that case, says Turgidson, a Soviet counterstrike would kill "only" 20 million Americans, whereas a Soviet first strike would kill 150 million.

And in the 1980s, several scientists, including astronomer Carl Sagan, introduced the idea of nuclear winter. They pointed out that the unbridled use of nuclear weapons would inject huge quantities of smoke and other particles into the Earth's atmosphere, blocking sunlight and lowering surface temperatures. Then, as happened with the asteroid strike that killed off the dinosaurs, plant life would be threatened, leading to massive starvation in addition to the other devastating nuclear effects. It's not clear that the calculations underlying these grim projections are fully accurate, but this scenario suggests a scientific basis for films like *On the Beach* that project a total loss of life after a nuclear exchange.

But the connection to science is much more tenuous in many films. Take the movie *Them!*, in which nuclear radiation in the New Mexico desert mutates ordinary ants into giant ones. It's true that among the harm radiation does to living things, it can damage the DNA in germ cells, that is, eggs or sperm. This mutation would not physically change the particular living organism but could be inherited in the next generation, as shown in experiment after experiment in plants and animals. And if the mutation could be carried through generations, you might think that the organism could indeed change radically—and for fast-breeding ants, rapidly as well.

Few random mutations are beneficial, however, which tends to make the organism that receives the mutated genes less likely to survive. Random mutations are also extremely unlikely to breed true, generation after generation, and very few mutations could be extensive enough to change an entire physical structure. There's another issue, too. Imagine you make an ant twice as big while keeping its bodily constitution the same. According to the mathematics of scaling, the bigger insect would weigh eight times more, but its legs would have strengthened only by a factor of four. The legs would carry twice the relative load for which nature had designed them. Every further increase in size would make the weight to strength ratio that much worse. At some point, the ant couldn't even stand up, unless the insect entirely altered its structure. But this would need a coordinated change in bodily properties far beyond what random mutations could induce.

Although radiation can't produce big ants, there is concern over its genetic effects on people. The children of atomic-bomb survivors have been carefully monitored for birth defects or changes in their DNA due to radiation-induced genetic damage in their parents. It's hard to separate genetic defects due to radiation from the many naturally occurring ones in humans, but the studies do not show significant damage carried from parent to child. Races of mutated beings arising in the aftermath of nuclear war, like the Screamers in *A Boy and His Dog*, are as unlikely as giant ants. (These genetic results don't mean that radiation doesn't harm children, along with everyone else. At Hiroshima and Nagasaki, pregnant women exposed to high radiation dosages gave birth to a larger than normal number of stillborn or mentally impaired children and children who died in their first year.)

Movies about nuclear warfare and nuclear power reflect realities of nuclear science and its consequences. But although the great majority of nuclear power plants operate without disasters, and radioactive sources are routinely used in medicine, films have painted nuclear fission as an entirely bad thing. Nuclear fusion is treated differently. While this is the process that underlies the most fearsome weapon we know, the hydrogen bomb, it's presented as potentially a good thing for humanity—potentially, since the controlled production of fusion power has yet to be achieved.

The fusion of hydrogen into helium keeps our sun going, and fusion reactions running on nothing but clean hydrogen could power our whole world, without using toxic radioactive fuel or generating pollution and global warming as fossil fuels do. Well aware of these benefits, scientists have been trying for a half century to make a fusion power source. This is like building an artificial star: to ignite fusion, the hydrogen (actually a mixture of two hydrogen isotopes, deuterium and tritium) must be heated to 100 million degrees Celsius and kept in a dense, confined state. This can, in principle, be done by using magnetic fields or titanic lasers. Enormous sums have been spent on huge fusion generators, but none has yet worked as a power source. In *Spider-Man 2*, though, Doctor Octavius comes close. Remarkably, after he ignites the reaction at high temperatures, it produces energy for a good fraction of a minute, longer than any real experiment has lasted, but then the reaction becomes unstable and destructive.

However, in 1989, electrochemists Stanley Pons and Martin Fleischman at the University of Utah seemed to skip right past the enormous difficulties of hot fusion. They announced the discovery of fusion at ordinary temperatures, cold fusion, using only a beaker of "heavy" water (in which the hydrogen in H_2O is replaced by deuterium) into which they dipped two metal electrodes that carried electric current. This simple apparatus, they reported, fused deuterium nuclei in the electrodes and produced energy. Other scientists rushed to duplicate the experiment, but no laboratory has ever been able to do so. Although a few researchers still believe in the feasibility of cold fusion, the majority have long dismissed it as mistaken.

Nevertheless, years after cold fusion came and went, it became central to the plot of *The Saint*. In her speech to science students, Emma Russell describes Pons and Fleischman as first acclaimed and then condemned. When a skeptical student comments that their experiment can't be replicated, she agrees that we don't know yet that cold fusion works; but she has a passionate belief that it *should* work because there it is in nature, just waiting to be harnessed (Emma's statement actually makes no sense, since as far as we know, nature supports only hot fusion in stars, not any kind of cold fusion). We soon learn that Emma is just the person to justify her own conviction, since her scientific research finally validates Pons and Fleischmann and shows that cold fusion is real and workable.

Chain Reaction also has something to do with cold fusion, or maybe it's hot fusion, or is it something else altogether? Dr. Barkley, head of the Hydrogen Energy Project, says there's enough energy in a glass of water to power Chicago for weeks but is coy about exactly how to get that energy out. To demonstrate hydrogen's energy potential, Barkley burns some along with oxygen. True, this is a relatively clean process, producing water as its residue, but there's no way Barkley should be touting burning hydrogen as a global energy source. That could hardly meet our needs—not to mention the fact that hydrogen mixed with air can explode. Later in the film, we see a madly bubbling container of water in the laboratory. This suggests cold fusion, but the scientists also use a laser and mention temperatures of millions of degrees, as if they're working on hot fusion. Which method finally wins the day? The movie gives not a clue. Even its title is confusing, since "chain reaction" is related to nuclear fission, not fusion.

Emma Russell in *The Saint* makes a breakthrough in a nonexistent scientific process, and in *Chain Reaction* we can't even tell how the power is allegedly extracted from hydrogen. But despite the shaky science, there is one important point about both films. After almost uniformly grim films about nuclear war, accident, and terrorism, these two stories end on an idealistic and uplifting note. Emma and Eddie both give away the limitless energy they've found to a world that badly needs it, cleansing our planet of toxic byproducts and addressing poverty. In a way, the gift of fusion power redeems the destructive events that began the nuclear age in 1945—and not only in the movies, but maybe someday in reality too. A consortium of nations, including the United States, plans to spend billions of dollars to build an experimental reactor in France by 2016, designed to prove that fusion power can be a commercial success.

Redemption is important in a world once dominated by a breakthrough in nuclear physics: the discovery of fission, applied to war and destruction. Now new breakthroughs come in molecular biology and biomedical science, with applications in genetic engineering. Through these, we better understand biological function and are finding new medical cures. But biomedicine also supports bioterrorism and biowarfare and introduces the possibility of radically changing ourselves, with unpredictable outcomes. Will the new biology bring humanity nothing but good, or will it some day require its own redemption? In the next chapter, we'll see what motion pictures have to say about this science and its consequences.

Chapter 6

Genes and Germs Gone Bad

> **Dr. Ian Malcolm:** The lack of humility before nature that's being displayed here . . . staggers me . . . your scientists were so preoccupied with whether or not they could, they didn't stop to think if they should.
>
> —*Jurassic Park* (1993)

> **Casey Schuler:** I hate this bug.
> **Colonel Sam Daniels:** Oh, come on, Casey. You have to admire its simplicity. It's one billionth our size and it's beating us.
>
> —*Outbreak* (1995)

> **Vincent Freeman:** I'll never understand what possessed my mother to put her faith in God's hands rather than those of her local geneticist.
>
> —*Gattaca* (1997)

In the 1940s and 1950s, physics was riding high. Its military contributions during World War II, such as the atomic bomb and radar, helped the United States and its allies defeat the Axis powers. When peace came, physics and the technology it supported entered people's lives. Today, this legacy is woven deeply into the fabric of society: every NASA mission, every nuclear power plant, every iPod, cellular telephone, and big screen TV owes its existence to physics and its applications. From space travel to the glossy electronic instruments seen onscreen in science laboratories, the same legacy is apparent in the movies.

But during that same post–World War II era, in 1953, another achievement came along that changed the balance and moved a different kind of science into prominence, in reality and in films. This was the determination of the structure of the DNA molecule—the double helix at the heart of the genetic code—by James Watson and Francis Crick with Maurice Wilkins and Rosalind Franklin. The work on DNA led to a Nobel prize for Watson, Crick, and Wilkins and brought biology into a new era.

Once, biologists could study only whole organisms and their gross features; that is, until the microscope was invented around 1600. Then biological research entered a tinier world, giving such insights as the germ theory of disease and the discovery of the cell as the smallest living unit. The study of DNA at the next level down, at the molecular scale, opened up an even smaller world and solved a very old puzzle. Through the mechanism of the genetic code, it explained how the characteristics of organisms are passed on, but also vary, from generation to generation. This was the beginning of molecular biology.

The ability to examine life in this fundamental way has provided new medical tools and remarkable new techniques of genetic engineering. In the past, we could modify plants and animals only through methods such as cross-breeding and selective breeding. Now, we can manipulate DNA at the microscopic level to add desirable characteristics or remove undesirable ones from an organism. We can even clone an organism, that is, make a genetic copy of it.

Genetic engineering and cloning have the potential to cure disease and to transform humanity and how we regard one another, although it's far from clear that all these changes are desirable. What is certainly unwelcome, but a fact of modern life, is that biotechnology also makes it easier to use diseases as weapons, with devastating possibilities for warfare and terrorism. As another fact of modern life, today's global connectedness brings newly frightening possibilities for pandemics, worldwide epidemics.

Between their scientific weight and their importance for the human condition, modern genetics and biomedicine have moved front and center in the attention of scientists and the public. The new biological science appears in films as well, but although the discovery of the double helix is a scientific drama laced with personality and conflict, it was carried out in quiet laboratories, not with the world-shattering impact of the atomic bomb. No massive event like the Manhattan Project or Hiroshima defines the start of the DNA era, and there's no major Hollywood film to tell that story (although it was told in the BBC television production *Life Story* [1987]).

Still, story lines about genetics and biomedicine powerfully merge science and human interest. They play on old threads in our psyches, such as the desire to make ourselves healthier, stronger, and smarter. Before we knew the genetic code, such tinkering with biological constraints could only be expressed through the repellent practice of selectively breeding people. The ancient Spartans left retarded or deformed infants to die; the Nazis favored a eugenics that would destroy "inferior" races while breeding greater numbers of the "master" race; and eugenics movements arose in other societies as well. Now deep changes in humanity seem eminently possible through genetic engineering, as explored in films like *The Sixth Day* (2000) and *Gattaca* (1997). Even *The Island of Dr. Moreau* (1996), the latest film version of H. G. Wells's 1896 story (*Island of Lost Souls* [1933] and *The Island of Dr. Moreau* [1977] were earlier versions), gives Moreau modern genetic tools to create a perfect race, a far cry from the brutal surgical methods Wells proposed in his day.

Another old theme in the human psyche that is amplified today is the dread of epidemics like the fearsome Black Death or bubonic plague of the Middle Ages, the subject of *Panic in the Streets* (1950). Later films such as *The Satan Bug* (1965) and *Outbreak* (1995) raise similar fears about contemporary diseases generated in military laboratories or arising in nature.

Panic in the Streets (1950) was directed by Elia Kazan and won an Academy Award for Best Motion Picture Story. It describes a modern emergence of the Black Death, which is transmitted by fleas from infected rats and wiped out at least a quarter of Europe's population in the mid-fourteenth century. In the story, the disease first appears stealthily at a late-night poker game near the New Orleans docks, where Kochak (Lewis Charles), a newly arrived illegal immigrant, feels ill. He leaves without giving Blackie (Jack Palance), a hoodlum, a chance to win back his money, so Blackie and two sidekicks (one played by Zero Mostel) follow Kochak and murder him, with Blackie pocketing Kochak's winnings.

The next day, Dr. Clinton Reed (Richard Widmark) of the U.S. Public Health Service finds that Kochak was carrying pneumonic plague, the form that is easily spread. Reed fears that the murderers are infected and will quickly distribute the disease, as once happened in Los Angeles, where dozens of people died soon after contact with an infected woman. A plague carrier could end up anywhere in the world in hours or days, he points out, and the killers must be found within forty-eight hours. Since Reed had Kochak's body cremated as a precaution, clues are scanty, but working with police captain Tom Warren (Paul Douglas), he tracks down the murderers in time and saves the city and country from the disease.

Fifteen years later, fears of new diseases rather than old ones took over when *The Satan Bug* (1965) brought germ warfare to the screen. Released during the Cold War, the film dramatically showed that biological war could be as terrible as nuclear war. The movie is based on a book by Alistair MacLean (written under the pen name Ian Stuart), and one of its screenwriters was Edward Anhalt, who, with his wife Edna, wrote *Panic in the Streets.*

The story begins in Station 3, a secret laboratory in the California desert, where a scientist has been murdered, and flasks containing the Satan Bug have been stolen. The Bug is a virus that could wipe out all life on Earth. Lee Barrett (George Maharis), a former government agent, is called in to investigate. He finds that the person behind the plot is Gregor Hoffman (Richard Basehart), who posed as a scientist to gain access to the laboratory and who wants to stop biowarfare research. Hoffman is more than intense: he's insane, and he threatens to release the Satan Bug in Los Angeles unless Station 3 is closed down. To show he's serious, he wipes out a small Florida town with the Bug, but Barrett finds him before he can infect Los Angeles, and the world is saved during a climactic battle on a helicopter.

Outbreak (1995) also describes governmental involvement with biowarfare. In 1967, a virulent disease flares up in a military mercenary camp in the African country of Zaire (now the Democratic Republic of the Congo). It's so deadly that U.S. Army medical officer Donald McClintock (Donald Sutherland) orders the camp bombed to keep the disease from spreading. Thirty years later, a village in the same Motaba region suffers another outbreak, and Col. Sam Daniels (Dustin Hoffman) of the U.S. Army Research Institute of Infectious Diseases is sent to investigate. He finds the disease to be 100 percent fatal, killing horribly within days by liquefying the internal organs. Back at USARMIID, Sam tells his boss, General Billy Ford (Morgan Freeman), that he's not sure the Motaba virus has been contained. Later we see a conversation between Ford and his superior officer, the same McClintock who called in an air strike in 1967, which suggests they know more than they've told Sam.

In fact, the disease has not been completely contained; it's being carried to the U.S. West Coast by a monkey aboard a freighter. When the creature arrives,

Jimbo Scott (Patrick Dempsey) smuggles it out of the animal quarantine facility where he works. After failing to sell it to a pet shop in the town of Cedar Creek, California, he frees the monkey in the woods, but not before he and the shop owner are infected. Soon townspeople are showing severe symptoms and dying, and the army moves in to quarantine Cedar Creek. Among those who fall ill is Sam's former wife, Dr. Robby Keough (Renee Russo), who arrived in town leading a team from the Centers for Disease Control.

Disobeying orders, Sam flies to Cedar Creek, where he finds that the virus is worse than in 1967; it has mutated and can now be transmitted through the air. Suspicious of Billy Ford and McClintock, he also finds that the army has been secretly turning the virus into a bioweapon and even has an antidote. Sam confronts Billy, who says he did what was necessary to make the perfect weapon. But now that the virus has changed, the antidote is no good. McClintock

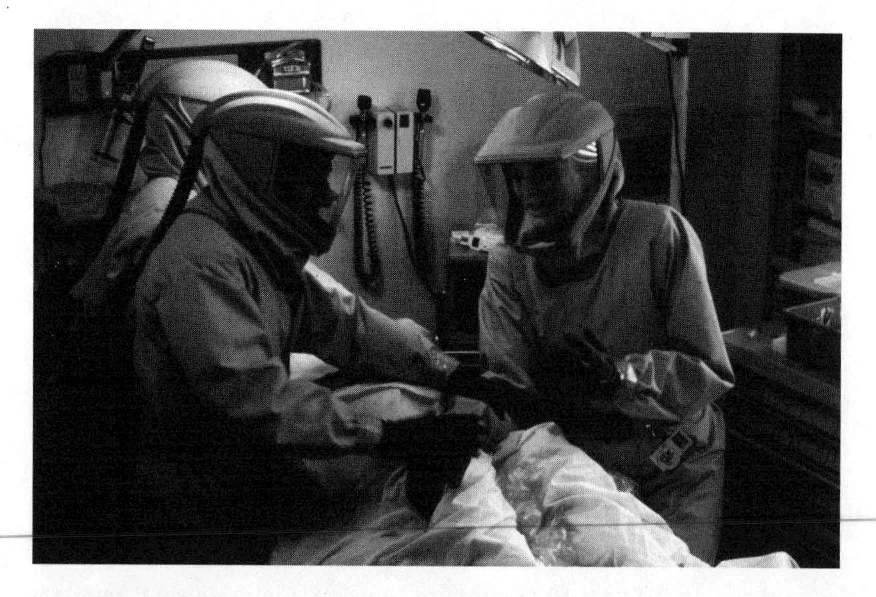

Outbreak (1995). Clad in protective gear, medical researchers Sam Daniels (Dustin Hoffman, left) and his former wife, Robby Keough (Rene Russo), work to contain the spread of the deadly Motaba virus (modeled on the real Ebola virus). The outbreak has been abetted by the secret efforts of military scientists to turn the virus into a weapon.

Source: Warner Bros/The Kobal Collection/Peter Sorel.

convinces the U.S. president that the only way to keep the virus from infecting the entire country is to destroy the town and its people, incidentally covering up McClintock's involvement.

To prevent this, Sam desperately searches for the monkey, whose antibodies can provide a serum against the virus. With Major Salt (Cuba Gooding) from his medical team, he commandeers a helicopter and finds the animal with the help of a little girl who had befriended it. McClintock tries to shoot down the helicopter, but Sam and Salt get the monkey back to Cedar Creek and start making serum. Nevertheless, McClintock orders that the town be bombed. Sam puts his helicopter in the path of the approaching bomber, which swerves and drops the bomb harmlessly into the ocean. The town is saved, the virus is stopped, and McClintock is arrested, while Robby recovers and seems ready to reunite with Sam as the film ends.

The diseases in these films—a medieval pestilence appearing in modern times, a vicious artificial disease, a murderous natural illness cultivated as a weapon within the U.S. military—may seem overblown or imaginary, but the medical science in *Panic in the Streets* and *Outbreak* has some firm connections to reality.

Far from being over and done with since the Dark Ages, the Black Death still had power when *Panic in the Streets* was released in 1950. Since then, it has been considered as a bioweapon, and even today, up to 3,000 cases of plague a year are reported worldwide. In 1994, the disease broke out in India, causing widespread panic and dozens of deaths. Noting how easy it is for disease to spread globally, the Centers for Disease Control immediately put in place surveillance to prevent importation of plague to the United States.

Although the last sizable plague outbreak on U.S. soil occurred in Los Angeles in 1924 (as Dr. Reed relates in *Panic in the Streets*), in the western United States the disease is still carried by fleas living on rodents such as prairie dogs. The United States typically has only about a dozen cases a year, but this old disease can still bite. In 2002, a couple visiting New York City became ill with plague contracted in New Mexico. Despite treatment with antibiotics, both the husband's legs had to be amputated below the knee, though the couple survived. In 2006, a woman in Los Angeles was diagnosed with plague and hospitalized, the first case there in twenty years—and unusual in that it occurred in an urban area.

Outbreak doesn't present a real disease, but its horrific Motaba virus is modeled on an actual disease agent, the Ebola virus. The image of the twisted, somehow evil looking virus shown in the movie is actually a picture of Ebola. As far as we know, this disease first broke out in 1976, in Zaire—as the movie shows for the Motaba virus—when it killed more than 300 people, and it has reappeared sporadically in Africa, most recently in 2003 and 2004. Between outbreaks, Ebola is thought to reside in a natural reservoir, an animal host (the species hasn't been identified, though there is some evidence that it's the fruit bat).

Humans are believed to contract Ebola through direct contact with an infected animal; the film fictionalizes this hypothesis, showing Jimbo Scott becoming infected when his smuggled monkey spits at him. Another connection is that Motaba is identified in the film as a hemorrhagic fever virus (HFV), a group of virus types that cause internal bleeding and to which Ebola belongs. However, although Ebola is deadly serious, the film exaggerates the effects of a hemorrhagic virus when it describes Motaba as "liquefying" bodily organs. An Ebola victim may bleed from his bodily orifices, but not with dramatic gushes of blood as if his insides were dissolving.

Ebola and plague are both prime candidates for biological weapons. In its bubonic form, plague produces painfully swollen lymph glands called "bubos" and is 50 to 70 percent fatal if untreated. Its more dangerous pneumonic form infects the lungs and spreads as easily as an ordinary cold. This form kills 95 to 100 percent of infected people if not treated within twenty-four hours, so Dr. Reed's haste in *Panic in the Streets* is justified. If the infection enters the blood, the result is a deep purple-black discoloration of the skin, hence the name Black Death. At this stage, the disease is absolutely 100 percent fatal if untreated. Even modern antibiotics don't guarantee recovery, as illustrated by the case in New York City.

According to legend, the Mongol hordes used plague as a bioweapon in 1346, catapulting the bodies of plague victims over the walls as they attacked the coastal city of Kaffa (now Feodosiya) on the Crimean Peninsula—although more probably, rats carried the disease into the city. In any case, ships that left Kaffa for Italy may have carried the Black Death into Europe. Before and during World War II, the Japanese military attempted to sow plague in China using infected fleas and was planning to aim the disease at the U.S. West

Coast just before the Japanese surrender in 1945. The Nazis also worked on methods to disseminate plague.

With no known treatment, and a mortality rate of 50 to 90 percent, Ebola, along with other HFVs, is listed by the CDC as a "category A" biological weapon that could be extremely difficult to keep under control. Most evidence indicates that Ebola is transmitted among humans by direct contact with an infected person or his bodily fluids, which means that the spread of the disease can be limited with appropriate precautions for patients and health personnel. However, there is some evidence that the virus can be transmitted through the air. In *Outbreak*, the really bad news comes when Sam discovers that the fictional Motaba virus can be airborne; similarly, airborne Ebola would make a truly fearsome weapon.

During the Cold War, the United States and the Soviet Union worked to put plague and HFV into aerosol form to enhance their deadliness. The Soviets "weaponized" plague and the Marburg virus, another incurable HFV—that is, produced them in forms that could be disseminated in bombs and missiles—and the United States also weaponized some diseases. In 1969, however, President Richard Nixon ordered that U.S. development of offensive bioweapons be halted and that existing stockpiles be destroyed. In *Outbreak*, McClintock's plot to weaponize Motaba violates this policy, and his arrest at the end of the film is long overdue. USARMIID, where Sam, Billy, and McClintock work in the movie, is a real agency at Ft. Detrick, Maryland, where researchers wear biohazard spacesuits to investigate dangerous diseases, but for defensive purposes only.

Defenses against bioweapons are necessary despite the 1972 Biological and Toxin Weapons Convention, which bans offensive use of biological agents. It was ratified by more than 150 nations, including the United States and the Soviet Union, but it lacks verification procedures. The Soviet Union and then Russia continued to weaponize diseases into the 1990s (so did Iraq, which didn't ratify the agreement until 1991, under pressure from the UN), and stockpiles of disease agents remain throughout the former Soviet Union. Now the concern is that rogue nations could launch biological attacks or that terrorists could steal disease agents from the Russian stockpiles, which are not well secured. To show the lengths to which terrorists will go, the Japanese cult Aum Shinrikyo—which killed and injured numbers of innocent people in 1995 when it released sarin nerve gas in the Tokyo subways—sent a group to (then) Zaire in 1992, apparently to collect Ebola virus for terrorist activities.

Bioweaponry represents a purely dark side of biological knowledge. Good and evil are knottier to sort out for genetic engineering, and especially for cloning. This is accomplished by inserting the nucleus of a cell from the donor, which contains the donor's DNA, into an egg that has had its DNA removed. The egg develops into an embryo and eventually a baby carrying the donor's genetic complement. The first animal cloned using cells from an adult donor (others had been cloned using cells from embryos) was the sheep Dolly. In 1997 Ian Wilmut and Keith Campbell, of the Rosling Institute in Scotland, created a worldwide sensation when they announced Dolly's 1996 birth. Much earlier, though, *The Boys from Brazil* (1978, from the book by Ira Levin), and *The Clonus Horror* (1979) (also known as *Parts: The Clonus Horror* or *Clonus*), were already taking genetic technology further. Later, *The Island of Dr. Moreau*, *The Sixth Day*, *Gattaca*, *Code 46* (2003), and *The Island* (2005) were still ahead of their time for human genetic manipulation, as was *Jurassic Park* (1993) for animal cloning, though not always with accurate science.

In *The Boys from Brazil*, Nazi hunter Ezra Lieberman (Laurence Olivier, who modeled his performance on real-life Nazi hunter Simon Wiesenthal) learns of strange activities among Nazis living in Paraguay long after World War II. They include Dr. Josef Mengele (Gregory Peck), the notorious Nazi physician called the Angel of Death. He carried out murderously sadistic medical experiments and primitive genetics studies involving twins on inmates of the Auschwitz concentration camp. In real life, he escaped from Germany to Argentina after World War II and finally drowned in Brazil in 1979, having never been apprehended.

In the movie, Lieberman finds that Mengele has ordered the murders of ninety-four men around the world, who share some peculiar characteristics: each is a sixty-five-year-old civil servant who, with a younger wife, has adopted a son with black hair, blue eyes, and pale skin. What soon emerges is Mengele's plot to create a Fourth Reich under a new Adolf Hitler. From embryologist Professor Bruckner (Bruno Ganz), Lieberman learns about cloning and realizes that Mengele has made multiple genetic copies of Hitler, but this isn't enough. The would-be Hitlers must also be raised in the same environment as the original, which includes the death of his father at age sixty-five.

Mengele selects the adoptive families and murders the fathers to match these conditions, increasing the chances that the clones will develop as Hitler did.

Lieberman and Mengele confront each other at the home of one of the adoptive fathers, racist and anti-Semitic Henry Wheelock (John Dehner). Mengele shoots Wheelock, and rants to his son Bobby (Jeremy Black, who also played three other clones) about his future as a new Hitler. Though Bobby, like all the clones, is icy and arrogant, he finds Mengele "weird" and sets Wheelock's vicious Dobermans on the Nazi, who's bitten to death. Lieberman is also shot, but survives and takes Mengele's list of clones. A member of a militant Jewish group asks Lieberman for the list so the young Hitlers can be killed, but Lieberman says this would be a slaughter of innocents and burns the list. The final scene, however, is ambiguous about where Bobby's genes and upbringing will take him.

The Island of Dr. Moreau portrays genetic alteration rather than cloning. It begins with U.N. diplomat Edward Douglas (David Thewlis) adrift in the Java Sea after his airplane crashes. Aboard the sailing vessel that picks him up is Montgomery (Val Kilmer), who takes him to a private island belonging to Dr. Moreau (Marlon Brando), a Nobel laureate geneticist. As Douglas learns from Montgomery and Moreau's beautiful daughter, Aissa (Fairusa Balk), Moreau has built a laboratory on the island to carry out experiments without interference from animal-rights activists.

We soon see that yes, both animal- and human-rights activists just might be concerned: Moreau has produced creatures who walk upright and talk, like humans, but have beastlike heads. They call Moreau—who looks pope-like, with his vast bulk clad in white—"Father" and consider him a god who gave them the Law, which absolutely bans killing. Horrified, Douglas confronts Moreau, who explains that the Beast People come from a fusion of animal and human genes. Douglas calls this satanic, but Moreau says he's seen the real devil in his microscope; it's "nothing more than a tiresome collection of genes" that controls the destructive part of human nature. Moreau airily dismisses the disfigurements that result from his genetic experiments, saying they're just stages in the process of creating a perfected race.

As Douglas realizes that Moreau and Montgomery are insane, events spin out of control. The Beast People rise up, kill Moreau, Aissa, and one another, and wreck the laboratory as they regress toward total beasthood. Douglas decides to take his chances by escaping on a raft, but he promises to bring back scientists to reverse the regression; however, the Sayer of the Law (Ron Perlman) who leads the Beast People, says, "No more science, no more laboratories, no more experiments. We have to be what we are . . . not what the Father tried to make us." Douglas survives, and later muses that sometimes he thinks the half-beast creatures Moreau created are not so different from the combination of animal and man in his fellow humans.

Gattaca continues the theme, imagining our society as it might be in the "not too distant future" when genetic manipulation is routine. Parents can choose the qualities they want their babies to have, selecting gender and other features from eye color to mathematical ability; adults can read one another's DNA from the saliva on a lover's lips or the perspiration on a doorknob. The result is a stratified social order: although it's illegal to engage in genetic

Gattaca (1997). In a high-tech future society that worships genetic perfection, routine DNA testing uncovers "in-valid" or "de-gene-erate" citizens like genetically flawed Vincent Freeman (Ethan Hawke). But illustrating that genes alone do not a person make, and with a little help from his friends, Vincent still manages to realize his life's dream, travel into space. *A Golden Eagle; see chapter 9.*

Source: Columbia/The Kobal Collection.

discrimination, the reality is that the plum positions go to the Valids, with impeccable genetic credentials, whereas the In-Valids, also known as "God children" or "de-gene-rates," are the world's janitors and toilet cleaners.

Young Vincent Freeman (Ethan Hawke) has a burning desire to break through this caste system. From childhood, he's wanted to go into space, but his parents didn't use a geneticist and he was born with a bad heart and myopia. His drive, though, animates him beyond his physical heritage. His genetically superior brother Anton (William Lee Scott) regularly outswims him, but the day comes when Vincent, with absolutely focused fortitude and determination, beats his brother and then saves him from drowning. As Vincent explains: "You want to know how I did it? This is how I did it, Anton: I never saved anything for the swim back."

The closest Vincent can get to space is janitorial work at Gattaca, a corporation that launches manned space flights. (The name "Gattaca" suggests DNA. The genetic information in DNA is coded in four compounds called bases: adenine, cytosine, guanine, and thymine. A readout of anyone's genetic code is a long string of the initials A, C, G, and T—the letters in "Gattaca"—arranged in an order unique to that person.) But then Vincent finds a way to get accepted by Gattaca at a higher level. Through a shadowy DNA broker (Tony Shalhoub), he meets Jerome Morrow (Jude Law). Jerome has a superb pedigree—as the broker says, "You could go anywhere with this guy's helix tucked under your arm"—and was competing as an Olympic swimmer until an accident left him in a wheelchair.

Vincent becomes Jerome, installing contact lenses to correct his vision and eye color, enduring a painful leg-lengthening operation, and using a urine sample from Jerome to take on his superior genetic identity. Vincent is immediately hired by Gattaca, and with Jerome supplying further urine, blood, and hair samples, continues passing as him. Soon Vincent is selected for a mission to Saturn's moon Titan, and he also connects with beautiful coworker Irene Cassini (Uma Thurman), who is also genetically imperfect.

But Vincent's plan is put in jeopardy when a Gattaca executive is murdered and the police find an eyelash identified as Vincent's. Still, he manages to maintain his false identity until he's about to board the spacecraft, when one last unexpected urine test reveals who he really is. However, Dr. Lamar (Xander Berkeley), in charge of testing, had long ago seen through Vincent's masquerade. Lamar is sympathetic because his own son didn't turn out genetically perfect, and he confirms Vincent as Jerome. As Vincent's spacecraft lifts off with engines firing fiercely, the true Jerome crawls into an incinerator, clutching his Olympic medal, and turns on the incinerator's own fiery blast—for as he once told Vincent, "I got the better end of the deal. I only lent you my body. You lent me your dream." The movie ends on this dual note, with Vincent's dream fulfilled as it inspires others—Jerome, Lamar, and Irene.

Code 46, a love story, also takes place in a future where genetic testing is woven into the fabric of society. Cloning and in vitro fertilization have become so common that a law, Code 46, has been passed with criminal penalties to "prevent any accidental or deliberate genetically incestuous reproduction." Prospective parents must be genetically screened; if their genetic makeups are too similar, conception is not allowed to proceed, and unsanctioned pregnancies must be aborted. People are required to carry papeles, documents with genetic information that divide the world into wealthy areas such as Shanghai, and third-world areas full of the dispossessed. Against this background, married insurance investigator William Geld (Tim Robbins) falls in love with laboratory worker Maria Gonzales (Samantha Morton), who has been forging papeles. But the government ends their relationship because it violates Code 46, and the film ends on a despairing note.

The Sixth Day (2000) also imagines a near future, "sooner than you think," where genetic engineering has gone even beyond its role in *Gattaca* and *Code 46*. After a momentous quote from Genesis—"God created man in His own image, and behold, it was good. And the evening and the morning were the sixth day"—the film continues with a short history of genetic achievement, shading from real events—1997, Dolly is cloned; 2000, the Human Genome project analyzes the human genetic blueprint—into society's imagined reactions: protests in Rome against human cloning; a failed human cloning experiment, with the clone destroyed by order of the Supreme Court; and finally the passing of so-called Sixth Day Laws that make it a major crime to clone a human.

But human cloning technology exists, as Adam Gibson (Arnold Schwarzenegger) finds out. Adam is no geneticist; he owns a helicopter charter business. On one particular day, life is good—he loves to fly, he loves his wife Natalie (Wendy Crewson) and young daughter Clara (Taylor Ann Reid), and it's his birthday. The only bad news is that Clara's beloved pooch Oliver has unexpectedly died.

That's where cloning comes in: Oliver can be exactly replaced by RePet, a company that stocks "blanks," animal embryos stripped of all DNA. When DNA extracted from a departed pet is inserted into a blank, that produces a physical duplicate. To make it behave just like the original, the technique of cerebral syncording is used to capture the dead pet's instincts and memories. These are downloaded into the blank via the optic nerve—and voilà, in just a few hours, it's as if doggy old Oliver (or for that matter, Fluffy the cat) had never died.

Adam is uneasy about cloning, even for a pet, but his main concern is his charter to take ultrawealthy businessman and entrepreneur Michael Drucker (Tony Goldwyn) snowboarding. At the last minute, Adam lets his partner Hank Morgan (Michael Rapaport) pilot Drucker under Adam's name. All seems to go well, until Adam returns home that evening and hears the sounds of a surprise birthday party. He looks through the window and sees himself—an exact double in appearance, voice, and manner—enjoying the party and a kiss from his wife.

Before a bewildered Adam can react, armed strangers appear and try to kill him, but he escapes and spends much of the movie finding out that Drucker is behind it all. He owns RePet, and also an organ transplant facility under chief scientist Dr. Griffin Weir (Robert Duvall). Weir's laboratory, however, secretly clones humans, using body blanks and syncordings just as RePet does.

Drucker wants to legalize human cloning because he himself is a clone, having once been killed and reanimated by cloning. On the day Drucker went snowboarding, he was again killed, by an anti-cloning fundamentalist who also killed Hank. Weir clones Drucker, and to cover up this crime, Drucker has Adam cloned in the belief that he's replacing Hank. When he discovers his mistake, he sends a team to kill one of the two Adams.

Once Adam gets this straight, he enlists his clone to bring down Drucker. In the climactic scenes, the two Adams use one of their helicopters to kill Drucker and yet another Drucker clone, and destroy Drucker's headquarters. Later, when Adam 1 compares notes with Adam 2, the clone worries that he's not completely human, but Adam 1 points out that his DNA is entirely normal. They resolve their duality by sending Adam 2 to far away Patagonia to open a second helicopter charter business. As the ship with Adam 2 steams out of the harbor, the original Adam salutes the clone with a helicopter flyby.

Another take on cloning appeared in *The Island*, although the same theme was expressed much earlier in *The Clonus Horror. The Island* is set in 2019, where people wearing spiffy but uniformlike white track suits live in pristine, sealed, and monitored high-tech surroundings. They're told that they've survived an environmental catastrophe that poisoned the world and are now protected in their own snug bubble. But they dream of the Island, the only unspoiled, beautiful spot left on earth. Periodically, they gather to hear the results of a lottery that will send one of them to this paradise.

Life in these mall-like surroundings is pleasantly bland, until Lincoln Six Echo (Ewan McGregor), who's the curious type, starts noticing little things that seem wrong, like a moth in the supposedly sterile environment. He gets some answers from McCord (Steve Buscemi), the engineer-cum-maintenance man who keeps things running, and then is shocked to find that instead of

The Island (2005). Jordan Two Delta (Scarlett Johansson) is one of many similarly dressed and cultivated clones raised in a sealed environment. Their fate is to be cruelly murdered on the operating table to provide new body parts for the wealthy. In reality, scientists hope that so-called therapeutic cloning will some day provide new hearts or livers, but without killing any clones, who—if they ever come to exist—would be fully human.

Source: Dreamworks/Warner Bros/The Kobal Collection.

being sent to any paradisiacal island, the latest lottery winner is dragged off to an operating table.

Lincoln's friend Jordan Two Delta (Scarlett Johansson) has just been chosen for the Island, so Lincoln urgently breaks the bad news to her and they escape into an outside world that's running along just fine. They find McCord, who explains that there was no worldwide catastrophe. Instead, Lincoln, Jordan, and the others are copies of real people, each of whom has paid millions to be cloned as the ultimate insurance policy. When a clone is called to the Island, it really means he's condemned to death so his genetically matched liver or lungs or heart can be removed and transplanted into the clone's sponsor.

Digesting this, Lincoln and Jordan decide to go in search of their sponsors, who they believe will take action when they hear how the clones are treated. Meanwhile, we learn that Dr. Merrick (Sean Bean), who masterminds the cloning operation, has developed technology that brings the clones to adulthood within a year. To assuage any guilt among the sponsors, Merrick lies, assuring them that "in compliance with the eugenics laws of 2015," the clones are "maintained in a persistent vegetative state. They never achieve consciousness. They never think or suffer or feel pain, joy, love, hate. It's a product . . . in every way that matters, not human."

Evading a team of high-tech bounty hunters that Merrick puts on their trail, Lincoln and Jordan find Lincoln's sponsor, Tom Lincoln (Ewan McGregor again), in Los Angeles. Tom pretends to be sympathetic but betrays the two clones to Merrick. In the resulting action, the real Tom Lincoln is mistakenly shot in the belief that he's the clone. The clones Lincoln and Jordan could simply take over Tom's rich and successful life, but they can't forget the friends they've left behind. Pretending to be Tom, Lincoln kills Merrick and destroys the entire operation. The final scene shows Lincoln and Jordan approaching that beautiful island in an ultramodern yacht that Tom had designed and built for himself.

The Clonus Horror, which features the original use of the same theme, is set in contemporary California, where a group of good-looking people in t-shirts and shorts is kept contentedly ignorant and isolated in an area called Clonus. Some, however, are chosen to be sent to what is described as the wondrous, happy land of America. As in *The Island*, one of them, Richard (Timothy Donnelly), finds something inexplicable in his environment—this time, an empty can of Old Milwaukee beer—that eventually leads him to the truth: he and his friends are clones who provide genetically matched organs for powerful people. These include presidential candidate Jeffrey Knight (Peter Graves), who's already received a transplanted heart. Richard escapes with a videotape to show the world what's going on, but things end less happily than in *The Island*. He returns to Clonus, where he finds that his pretty blonde friend Lena (Paulette Breen) has been lobotomized to keep her quiet. The videotape does eventually surface in the real world, and we know that Knight and the Clonus operation will be brought down. But that's small consolation for Richard; the final shot in the film shows that his heart has been harvested.

Among these different cinematic predictions for genetic engineering, the most far-fetched is in *The Sixth Day*, with its idea of defeating death by replacing dead people with their identical living selves. This is presented as routine, a mere hiccup in life. One of Drucker's employees has been killed and cloned so often that she's utterly blasé about her latest return from the dead; her main concern is that she needs to have her ears pierced again!

But sadly for the film's premise, even if you accept the notion of somehow recording and downloading personality and memory—which no one in reality has the slightest idea how to do—cloning is not about Xeroxing an adult; all it does is create an embryo. If Drucker were cloned the instant he was shot, say at age forty, it would still take nine months plus forty years to produce another forty-year-old Drucker. Adam in *The Sixth Day* could never meet his clone of the same age. *The Clonus Horror* gets this basic point right, for along the way it implies that its clones must grow from babyhood to adulthood. *The Island* at least has the grace to make cloning less than instantaneous, since it takes a year for Dr. Merrick to bring a clone to adulthood. However, although clones are necessarily younger than their originals, the fact that your genetic twin will live on, more nearly "you" than your child, could confer a kind of life extension.

It shouldn't be surprising that these films push genetic engineering beyond reality. The stakes are high, and it's hard to avoid wishful thinking. That's shown in the history of cloning, which is longer than you might think and has had its share of hype and worse. In 1902, the German embryologist Hans Spemann, who was to win a Nobel Prize in 1935, divided a salamander embryo in two. Each half went on to grow into an adult salamander, showing that cells from an early stage embryo contain all the genetic information needed to create a new organism. In 1938 he proposed a "fantastical experiment" that would put the nucleus of a cell into an egg without a nucleus. This is nuclear transfer, the heart of cloning, expressed years before DNA had been discovered.

The process wasn't called "cloning" then: the name, from the Greek for "twig," was proposed by the famous British biologist J. B. S. Haldane in 1963. Haldane painted a rosy picture of cloning "from cells of persons of attested ability [which] might raise the possibilities of human achievement dramatically," especially, he thought, if a great artist or mathematician were to educate his or her own clones. But he also noted that as we learn from our mistakes, "we shall no doubt clone the wrong people," reproducing those like Hitler who should never have been born in the first place.

As genetic technology developed, it was sometimes difficult to separate valid results from sloppy research, exaggerations, and even hoaxes—and I'm talking real life, not the movies. By the 1970s, frogs and tadpoles had been cloned, but the cloning of mammals and people seemed a distant possibility. However, in a 1978 book entitled *In His Image*, David Rorvik tells a purportedly true story about a multimillionaire who cloned himself. This is like the plot of *The Sixth Day* or *The Island*, but the book was just as much a fiction as the films, for Rorvik had made it all up. Not long after, in 1981, researchers Karl Illmensee and Peter Hoppe claimed to have cloned the first mammals, three mice. These results came under fire, and though they were never proven fraudulent, the scientific community rejected them.

But in the later 1980s, mammals in the form of a sheep and cow were truly cloned, using cells from embryos. Then in 1996, Ian Wilmut and Keith Campbell made their major breakthrough: they cloned the sheep Dolly from cells taken from a mature ewe, although it took nearly 300 tries to yield one fetus that survived when implanted into a surrogate mother sheep. This put cloning on the map as a real possibility for humans, which inspired others, though not in good ways.

In 1998, Richard Seed, a Chicago physicist who had been involved in fertility research, received wide attention when he announced his intention to clone a human, but nothing came of the project. In an apparent hoax, a fringe religious group called the Raelians claimed in 2002 to have cloned a human but never presented an infant or any details for independent scrutiny. And in 2005, Korean scientist Hwang Woo Suk, who had apparently done stellar work on so-called therapeutic cloning, was found guilty of fabricating his results.

Therapeutic cloning is a medical treatment that would use cells from a patient's body to generate a genetically identical embryo, which would yield stem cells. These are the all-purpose cells that turn into whatever the embryo needs as it develops—blood, heart muscle, brain cells. The stem cells could be implanted in the patient's body to repair damage caused by illnesses such as heart disease. That's where cloning is important. Currently, a patient with an implant such as a transplanted kidney is condemned to a lifetime regimen of powerful antirejection drugs as his immune system tries to fight off the alien cells. Genetically matched stem cells would eliminate this. Should it also prove possible to grow the cells into complete replacement organs, as some researchers want to do, therapeutic cloning would also eliminate the ethical complications of human organ donation.

If therapeutic cloning is the basis of the organ-transplant facility Dr. Weir operates in *The Sixth Day*, this part of the story could come to pass as a benefit to humanity. *The Clonus Horror* and *The Island* also recognize the value of genetically matched organs, but in showing them as ripped from the bodies of cloned humans rather than grown externally from stem cells, they present a truly evil use of genetic engineering.

This is not the only moral question raised by these movies. In *Gattaca*, Vincent and Irene watch a classical pianist perform a work of great intricacy with total mastery and find that he's been genetically engineered to have twelve fingers. This makes him a superior pianist, but is it ethical to change him from the norm without his consent? And how far could such changes be pushed before the result is considered monstrous? The back story for *The Sixth Day* raises another question when it notes that an early, failed attempt at human cloning had to be "destroyed"—a euphemism for either abortion or murder, depending on whether the clone was a fetus or an infant. And *Mimic* (1997), although it's about genetically altered insects rather than humans, presents another warning about genetic manipulation. A new disease carried by cockroaches is killing hordes of children in Manhattan. Conventional medicine can't cure it, but entomologist Dr. Susan Tyler (Mira Sorvino) becomes a hero when she genetically engineers a race of insects that kill the roaches and stop the disease in its tracks. However, though Tyler designed the killer insects to die out after completing the job, they survive and evolve into a new threat that she and her colleagues barely manage to avert.

The horrible possibility of even well-meant genetic engineering going wrong is one reason that real human cloning has met considerable, sometimes visceral resistance. To some, cloning a human, or altering humans by tinkering with their genes, usurps God's prerogatives. To those who think an embryo is a human being, therapeutic cloning seems wrong as well, since the process destroys the embryo that yields the stem cells. After the news of Dolly's cloning in 1997, President Bill Clinton proposed that human cloning research be banned for five years, saying, "Any discovery that touches upon human creation is not simply a matter of scientific inquiry. It is a matter of morality and spirituality as well."

Although the science in *The Sixth Day* is over the top, the film does reflect these concerns: anticloning demonstrators block entry to Drucker's enterprises, chanting "God doesn't want you to go in there"; Adam worries that his daughter's cloned pet would lack a soul and therefore be dangerous; and Drucker's fundamentalist assassin calls him an "abomination to God" because he's a clone.

A confrontation between Drucker and Adam concisely lays out the issues, to some extent characterized as a conflict between science and religion. Drucker says the Sixth Day Laws have curtailed legitimate research in cloning, which would allow us to keep our Mozarts, our Martin Luther Kings. Adam replies:

> **ADAM:** And who decides who lives or dies? You?
>
> **DRUCKER:** You have a better idea?
>
> **ADAM:** Yes, what about God?
>
> **DRUCKER:** You're one of those. I suppose you think science is inherently evil.
>
> **ADAM:** I don't think science is evil. But I think you are.
>
> **DRUCKER:** If you believe God made man in His own image, you also believe He gave man the power to understand evolution, to exploit science, to manipulate the genetic code. I'm just taking over where God left off.
>
> **ADAM:** If you really believe that, you should clone yourself while you're still alive . . . so you can go f**k yourself.

These films also make points about the power of genes alone to define a person. In *The Boys from Brazil*, Mengele's plan to create new Hitlers recognizes that environment matters too (as the Nazis well understood. It wasn't enough to produce Aryan children; Nazi youth also had to be indoctrinated.) But in *Gattaca*, society has bought heavily into genetic determinism, the belief that you are only your genes in physical makeup and behavior. Vincent resists this, and his desire to go into space carries him to a level his society says he could not and should not achieve. This is where the situation in *Gattaca* stands in for real-life discrimination that defines a person purely by skin color, gender, or age.

Further, there is no such thing as complete genetic predictability. Part of what makes up any organism, including a human, is random genetic variation that can't be controlled; no matter how powerful the techniques of genetic engineering, complete mastery over every aspect of a person is forever beyond our reach, as some motion pictures illustrate. Despite the manipulations of genes and environment carried out by Dr. Merrick in *The Island* and by Dr. Moreau on his personal island, the clones and bioengineered creatures they create behave in unexpected ways.

Nor can we have foreknowledge of an individual's trajectory in life. While *The Sixth Day* pretends that cloning would allow us to defeat death, *Gattaca* pretends that genetic knowledge would make death more immediate by making it predictable. Mere seconds after Vincent's birth, a nurse reads his death sentence from his genetic code: he'll die of heart disease at the age of 30.2 years. But it's extremely unlikely that the most complete DNA data imaginable could predict any individual's life span; at most, it could give statistical probabilities of various diseases whose genetic roots are understood, without taking into account environmental factors, random variations, and medical advances.

The genetic engineering presented in these films depends on knowing the human genome; that is, all the DNA in a human, three billion pairs of the four bases G, A, T, and C packaged into chromosomes. Humans have twenty-three pairs of chromosomes in each cell (one half of each pair from the mother, one from the father). These contain about 30,000 genes, particular strings of DNA that carry the information needed to make specific proteins, the body's building blocks. A good deal about these genes has been learned through the Human Genome Project, which began in 1990 and released its results in 2001. Our genome has been sequenced, meaning the order of all the bases has been determined, and is now being mapped, meaning the locations of the genes on the chromosomes are being determined.

The next phase, also underway, is to determine which genes connect with which bodily functions and properties and with which diseases. That's the necessary background for genetic manipulation that would change a person's makeup or eliminate susceptibility to certain diseases. We're not yet tailoring human genes to order, but a first step, testing for genetic indicators of disease, is well established and points to the moral quandaries to come.

About 700 clinical genetic tests are now in use. Some count the number of chromosomes. This detects Down's syndrome, which produces babies with mental retardation, distinguishing facial features, and heart defects, and is caused by an extra chromosome. Other tests seek specific faulty genes, such as two that are markers for breast cancer, or specific DNA sequences. For example, a person whose DNA shows a long repeated sequence CAGCAGCAG . . . on a particular chromosome is at risk for Huntington's disease, which causes nerve deterioration. Still other tests can predict whether a person will respond well to particular drugs for certain diseases.

Such early warnings can be a mixed blessing because our ability to identify these diseases outstrips our ability to cure them. A pregnant woman who faces a diagnosis of Down's syndrome in the fetus she carries can choose only to go through with the birth or abort it. For this reason, some people prefer not to undergo genetic testing. As cures for these diseases are developed, this issue will diminish, but there will always remain the question of deciding who should be genetically tested. And if genetic engineering reaches the point where it can produce a superior human, the burning questions will be: who's eligible for the improvement, and how should such people fit into society—the very situation painted in *Gattaca*.

Genetic engineering and human cloning may never come to pass, for reasons scientific, ethical, or both. Plants have been cloned, however, since the first cutting was taken and grew into a genetic copy of the original. Now we have genetically enhanced food plants and grains, such as a modified rice that carries increased levels of vitamin A and iron. Animal cloning, too, has its established uses. It has been valuable in efforts to develop genetic technology for humans, and increasingly, animals are cloned to produce commercial breeding stock. In late 2006, the U.S. Food and Drug Administration declared that meat and milk from most cloned animals was safe for human consumption. The FDA stopped short of approving the sale of food from cloned creatures, but that final approval may have come by the time you read this.

Though some people fear that food from cloned or genetically engineered animals—so-called Frankenfood—is unsafe to eat, the morality of cloning animals has never been in dispute in the same way as cloning people. In concentrating on animal cloning, *Jurassic Park* tells an imaginative story and presents the cloning process without raising these particular ethical issues, though it suggests others.

As the story begins, entrepreneur John Hammond (Richard Attenborough) is hearing from corporate lawyer Donald Gennaro (Martin Ferraro) that Hammond's brand new Jurassic Park needs scientific approval before investors can believe it's safe. The theme park, located on an isolated island, offers something unique: living dinosaurs, genetically reconstructed from DNA preserved from the time of dinosaurs, 65 million years ago and more. To check out the park, Hammond recruits paleontologist Dr. Alan Grant (Sam Neill); his colleague and girlfriend, paleobotanist Dr. Ellie Sattler (Laura Dern); and chaos theorist, or "chaotician," Dr. Ian Malcolm (Jeff Goldblum).

Arriving at the island, the scientists are awestruck to see an enormous brachiosaur, a dinosaur the size of a whale with a neck thirty feet long, calmly munching on tree tops, while Hammond beams with pride. Alan is stunned to find that the creature is warm blooded, upsetting years of theorizing about dinosaurs, and he nearly faints when Hammond announces that he also has a *Tyrannosaurus rex*, a T. rex.

Hammond shows how this was done through an animated cartoon in which Mr. DNA explains that DNA is the blueprint of a living creature and that the dinosaurs have left behind DNA samples. Mosquitoes that fed on dinosaur blood millions of years ago became trapped in sticky tree sap that hardened into amber. Hammond's scientists extract the dinosaur blood from mosquitoes and use supercomputers to examine the DNA it contains. Since the DNA is old, its genetic sequence has gaps, which are repaired with frog DNA. Then the complete DNA sequence is inserted into unfertilized ostrich or emu eggs. Alan, Ellie, and Ian watch as one of these eggs hatches into a baby velociraptor, the smartest, fastest, and most dangerous of all dinosaurs.

Though the visitors are impressed by this miraculous achievement, Ian is worried that the dinosaurs will breed in the wild. But Hammond has thought

Jurassic Park (1993). Dedicated paleobotanist Ellie Sattler (Laura Dern) and paleontologist Alan Grant (Sam Neill, foreground right), with young Tim Murphy (Joseph Mazzelo, left rear), minister to an ailing dinosaur cloned from DNA many millions of years old. *Jurassic Park* introduced astonishing special effects to display its life-like dinosaurs and helped make "cloning" a household word. *A Golden Eagle; see chapter 9.*

Source: Amblin/Universal/The Kobal Collection.

of this: his scientists create only females. Ian isn't convinced, saying, "If there's one thing that evolution has taught us, it's that life will not be contained. Life breaks free, it expands to new territories and crashes through barriers." And trouble is brewing elsewhere, too. Dennis Nedry (Wayne Knight), in charge of the computers that run Jurassic Park, including its electrical dinosaur-proof fences, has been offered $1.5 million by a competing group to steal dinosaur embryos from the park's laboratories.

When the scientists, Hammond's two grandchildren, and Gennaro go off in self-guided tour cars to see the dinosaurs, everything quickly goes wrong: computer glitches develop, Nedry shuts down the security systems so he can take the embryos, and a storm cuts electrical power—which allow the dinosaurs to go free. The T. rex eats Gennaro, and another dinosaur kills Nedry as he tries to flee the island. Alan and the children escape the T. rex; however,

Alan spots something ominous, newly hatched dinosaur eggs. Some frogs, it seems, can spontaneously change sex, and the frog DNA mixed with dinosaur DNA allows the female dinosaurs to do the same and breed.

Back in the Park's control room, Hammond and Ellie manage to restore the electricity and the computers. But soon Ellie, Alan, and the children find themselves hunted by two ferocious and intelligent velociraptors. Just as the humans seem hopelessly cornered, with the adults sheltering the children, the T. rex appears. Huge compared to the raptors, it gobbles them with ease. The adults and children run for a jeep driven by Hammond, and Alan says, "Mr. Hammond, after careful consideration, I've decided not to endorse your park." Hammond responds, "So have I."

The movie ends with all safely aboard a helicopter flying off into the sun and toward the mainland, as Hammond gazes ruefully at a mosquito in amber on the head of his walking stick. In the sequels *The Lost World: Jurassic Park* (1997) and *Jurassic Park III* (2001), the dinosaurs continue to breed, multiply, and evolve. The velociraptors prove themselves more intelligent than dolphins or primates and reach a point where they communicate with one another.

Between the instructional Mr. DNA cartoon, and the scene where a baby dinosaur breaks out of an egg, *Jurassic Park* correctly demonstrates the basics of cloning. But how about the idea of reproducing an enormous dinosaur from a tiny residue of DNA held within amber? Could nature really have arranged this charmingly convenient method of bringing us a living relic of the past?

At a quick glance, this seems reasonable. Although most prehistoric animal remnants are bones and teeth, amber can preserve soft animal (or plant) tissues. It can also trap insects from the era of the dinosaurs, such as 85-million-year-old mosquitoes, and a 100-million-year-old bee found in 2006. But although ancient embalmed insects exist, it seems that viable 65-million-year-old DNA doesn't. Over millions of years, the DNA deteriorates and becomes contaminated. While scientists aren't planning to build a Jurassic Park any time soon, they would love to find ancient DNA to study evolutionary processes. In the early 1990s, some scientists reported finding old DNA in amber. These results haven't been reproduced, however, and subsequent research has convinced the scientific community that they're erroneous, possibly arising from modern contamination. And anyway, if you could extract DNA from a mosquito, it's more likely to be mosquito DNA than dinosaur DNA.

The remaining steps in John Hammond's dinosaur renaissance are suspect as well. You would need a dinosaur egg, not an ostrich egg, to properly support the embryo—and with a living dinosaur egg, you wouldn't need to do any cloning in the first place. Even if a baby dinosaur were to hatch, no one knows its environmental and nutritional needs, which might include food long since vanished from our world. (Also, sad to say, the intelligence ascribed to the velociraptor dinosaurs in the *Jurassic Park* series is far greater than what scientists think this species really achieved, and there are other problems with how the dinosaurs are portrayed.)

But although there's little hope of building a *Tyrannosaurus rex* from a preserved mosquito, there are other possibilities to harvest old DNA. In 2005, David Penney, a paleontologist at the University of Manchester, examined amber from 15 to 20 million years ago and found droplets of spider blood that could hold uncontaminated spider DNA. Also in 2005, Mary Schweitzer of North Carolina State University and her colleagues reported extracting blood vessels from the skeleton of a T. rex held in the Museum of the Rockies in Montana. The vessels are still flexible and elastic after 68 million years, and, especially startling, they contain structures resembling cells and cell nuclei. Although more testing is needed, there may be a chance that unharmed DNA has survived and could be extracted from these.

Biology and biomedical science have never seemed as marvelous as they do now. From the wondrous, if remote possibility of extracting ancient species from preserved DNA to efforts to genetically bootstrap ourselves into something better, the story is more fascinating and optimistic than ever. There's a shadowy side to the biological revolution, too—deadly diseases bred in laboratories, cloning experiments gone wrong, debates over how far humanity should go in modifying and creating life. Films have shown all this, from Dr. Moreau's warped experimentation, to diseases made for warfare, to the genetically perfect but troubling world of *Gattaca*. This range shouldn't be surprising, since any technology made to benefit humanity can also undo humanity. If that lesson didn't come home with these films, it will in those we contemplate next, where computers and robots become our masters instead of our slaves.

Chapter 7

The Computers Take Over

Dr. Charles Forbin: I think *Frankenstein* ought to be required reading for all scientists.

—*Colossus: The Forbin Project* (1970)

Morpheus: The Matrix is . . . the world that has been pulled over your eyes to blind you from the truth.

—*The Matrix* (1999)

The First Law of Robotics: A robot may not injure a human being or, through inaction, allow a human being to come to harm.

—*I, Robot* (2004)

Starting in the 1920s, if you had viewed too many science fiction films, you might be forgiven for nightmares in which you were surrounded by mechanical creatures that could do anything a human could—walk, talk, think, fight—only better, and often with evil intent. Or if not walking, talking robots, then you might be terrified by massive machine brains that could outthink humans or could suck humanity into an imaginary virtual world; or by hybrid beings, loathsome combinations of warm-blooded human, cold mechanical device, and inscrutable electronic chip.

The creation of various forms of artificial or hybrid life and thought is a surprisingly old theme, dating back to the ancient Greeks. After this idea appeared in mythology and legend, it was expressed in books, and then in films about robots, androids (also robotic, but specifically made to look human), cyborgs (cybernetic organisms, usually a human brain in an artificial body), and computer minds or artificial intelligence (A.I.).

These movies have a long history, and by 1927, the futuristic film *Metropolis* featured a striking mechanical robot. Then in 1931, *Frankenstein*, based on Mary Shelley's 1818 book and starring Boris Karloff as Frankenstein's creature, set a standard for artificial creatures, from the vegetable monster in *The Thing from Another World*, to the moment in *Star Wars Episode III: Revenge of the Sith* (2005), when Darth Vader, freshly created as a cyborg from what's left of Anakin Skywalker, struggles upright from the operating table and lurches into his new life as an evil force.

In older legends and stories, too, artificial beings perform plenty of evil. The Golem of Prague, made out of clay by a rabbi to protect the sixteenth-century Jewish community of that city, turns on its maker and kills him. Frankenstein's creature, in the book and movie, murders innocent people. Now, although glossy high technology has replaced used body parts, movie plots remain the same: our own creations, robots or artificial minds, rise up to dominate and kill humans. In other scenarios, people are changed or debased by unholy alliances with artificial entities, or by falling under their control. Professor Frink, resident scientist on the popular TV show *The Simpsons*, puts it concisely in his Frink's Law: any and all robots will eventually turn on their human masters.

In fact, one of the earliest robots to appear in film is up to no good. That's the mechanical creature in the silent movie *Metropolis*, made by the great German director Fritz Lang. The futuristic city of Metropolis is filled with awe-inspiring skyscrapers inhabited by the wealthy and powerful, but below ground, hordes of downtrodden workers keep things running while enduring a squalid and regimented existence.

Metropolis's Master, Johann "Joh" Frederson (Alfred Abel), wants to keep the workers in their place, but Frederson's son Freder (Gustav Fröhlich) wants to help them after falling in love with one of their number, the good and beautiful Maria (teenage actress Brigitte Helm, in the role of a lifetime). Frederson plans to keep the laborers impotent through new technology developed by C. A. Rotwang (Rudolf Klein-Rogge), a kind of wizardly scientist who is making robots that could replace human workers.

Rotwang's prototype robot is both aggressively machinelike and recognizably womanly. Its massive metal joints wouldn't look out of place on a modern assembly line robot, but it also has breasts and hips and a provocative female face—an appearance that easily makes it, or her, the film's most memorable character.

Through an electrical transference process, Rotwang turns the robot into an android duplicate of Maria (also played by Brigitte Helm). The replica Maria has an evil streak and, unlike the saintly real Maria, radiates sexuality. It winks and leers and performs an erotic dance before Frederson and his powerful friends. Frederson sends this turncoat pseudo-Maria, whom the workers trust, to provoke dissension among them. In the remainder of the film, the workers rebel, but after other twists and turns, the story concludes with peace restored, the assurance of a better day for the workers, and a reunion between Freder and the real Maria.

The android Maria caused plenty of trouble but didn't kill anyone or try to take over the world. Robots, cyborgs, androids, and computers in later films do one, the other, or both. The most famous of these, from Stanley Kubrick's 1968 film *2001: A Space Odyssey* (based on Arthur C. Clarke's 1951 short story *The Sentinel*), is HAL, the A.I. that operates a spaceship carrying astronauts and scientists to the planet Jupiter. HAL is immensely capable. It makes major decisions for the mission and monitors everything aboard ship, with capacity left over to chat and play chess with the crew and make personable conversation during a television interview (HAL's voice is provided by Douglas Rain).

But perhaps as a result of too much intelligence, or simply a short circuit, HAL goes mad. It murders the scientists who are in deep sleep for the voyage, and one of the astronauts, leaving only Dave Bowman (Keir Dullea), who must disable HAL to survive. As Dave pulls HAL's circuit boards, diminishing its mind step by step, the A.I. displays humanlike reactions:

> I know I've made some very poor decisions recently, but I can give you my complete assurance that my work will be back to normal . . . will you stop, Dave . . . my mind is going . . . [slows down] . . . I'm afraid . . .

Finally HAL reverts to an earlier, childlike stage in its development and, like a five-year old, sings the song "Daisy, Daisy" with its especially meaningful line "I'm half-crazy."

HAL's murderous activities might be excused as the actions of a mentally ill A.I., but in *Blade Runner* (1982), the artificial beings—androids called replicants—are aggressively violent. The film, based on Philip K. Dick's novel *Do Androids Dream of Electric Sheep?*, takes place in Los Angeles in 2019. As in *Metropolis*, the future L.A. mixes tall, glossy towers with lower, darker, grittier levels where the streets are always slick with rain and neon signs glow day and night. Against this film noir–ish backdrop we follow the interaction between the rogue replicant Roy Batty (Rutger Hauer) and the human blade runner, or special policeman, Rick Deckard (Harrison Ford), assigned to destroy him.

Batty is the highest type of replicant, a Nexus 6 model made for combat, with physical skills and intelligence that have suited it to work in a human colony on

Blade Runner (1982). In a future world, Roy Batty (Rutger Hauer) is a powerful, intelligent replicant, or android, that has developed self-awareness and the knowledge that its days are numbered. Its ruthlessly violent behavior is mixed with moments of humanity and poetry. In reality, robotics technology is still far distant from creating fully capable, self-aware artificial beings. *A Golden Eagle; see chapter 9.*

Source: Ladd Company/Warner Bros/The Kobal Collection.

a distant planet. But like all replicants, it faces an absolute deadline. These androids have been designed by their creator, Eldon Tyrrell (Joe Turkel) of the Tyrrell Corporation, to last only four years; otherwise, the fear is that they will develop emotions and become difficult to control. Along with other replicants, Batty escapes from the space colony, murdering a human shuttle crew along the way, and returns to Earth to confront Tyrell before its time runs out.

The movie explores how artificiality and humanity mix. Deckard's job is to eliminate the androids, yet he falls in love and has sex with Rachael (Sean Young), a new advanced replicant model that believes itself to be human and is crushed when it finds that's not true. Batty also shows a mixture of humanity and inhumanity. It uses extreme violence and murder to reach its creator; yet when they meet, it's like a father and child reuniting. Tyrell calls Batty a prodigal son, and Batty bows its head in atonement for "questionable" acts in its past. But Tyrell can't or won't extend Batty's life, and an instant after kissing its "father," the android's remorseless side again emerges as it crushes Tyrell's head between its powerful hands.

Roy Batty kills its creator for a specific personal reason; it feels Tyrell has cheated it out of a full existence. In *The Terminator* (1984), the first of a series, the title character (Arnold Schwarzenegger) is a human-seeming android designed to pursue and murder whomever its programming dictates. It is the creature of conscious computers that dominate the future world of 2029. These machines trace their lineage back to computers originally built to defend the United States against its enemies and that later became conscious and decided to rid the world of humanity.

The few remaining human rebels in 2029 are led by one John Connor. To eliminate him, the intelligent computers send the Terminator back to 1984 Los Angeles to kill Connor's mother-to-be, Sarah Connor (Linda Hamilton), before she can give birth to her son. But the future humans also send back one of their own, Kyle Reese (Michael Biehne), to tell Sarah about the merciless model T-800 Terminator:

> Underneath, it's a hyper-alloy combat chassis—microprocessor-controlled, fully armored. Very tough. But outside, it's living human tissue—flesh, skin, hair, blood. . . . It can't be bargained with. . . . It doesn't feel pity, or remorse, or fear. And it absolutely will not stop, ever, until you are dead.

Kyle and Sarah keep a step ahead of the Terminator, but in the final scenes, it nearly carries out its mission, even after Kyle, who has fallen in love with Sarah, sacrifices himself to blow it to pieces. The Terminator's upper half survives and continues to pursue Sarah until she manages to pound it flat in a hydraulic press, which ends the movie.

There's more to come, though. Sarah is pregnant by Kyle, and the child she bears grows up to become John Connor. The computers in 2029 haven't given up on nipping him in the bud, and in the sequel, *Terminator 2: Judgment Day* (1991), they send an improved model T-1000 Terminator (Robert Patrick) back in time to kill John as a boy (Edward Furlong). Countering this, the human rebels send back their own Terminator (Arnold Schwarzenegger again), but this time, reprogrammed to act as John's bodyguard. This "good" Terminator represents a father figure to John, and it, Sarah (Linda Hamilton again), and the boy come together like a family. The next sequel, *Terminator 3: Rise of the Machines* (2003), also features this once bad, now good android again facing another evil Terminator, a female version (Kristanna Loken).

For all its machine ruthlessness, the Terminator is programmed to kill only one person at a time. *Stealth* (2005) introduces an A.I.-guided weapon that does far more harm. You may not have seen the film, which had only a brief theatrical run, but it makes interesting points about intelligent weaponry. It begins as three superb U.S. naval aviators—Lts. Ben Gannon (Josh Lucas), Henry Purcell (Jamie Foxx), and Kara Wade (Jessica Biel)—who fly advanced Talon combat aircraft off an aircraft carrier, are told by their commander, Capt. George Cummings (Sam Shepard), that a new wingman is coming aboard.

Soon a new aircraft appears and makes a perfect landing, but with no one at the controls. This warplane, with a streamlined yet hulking look, *is* the new wingman. It's controlled by an A.I. that uses quantum computing to handle data at the staggering rate of 4,000 gigabytes a second. Eddie, as the aircraft is nicknamed (from EDI, Extreme Deep Invader), understands human speech and responds in an affectless voice like HAL's. It has flashes of personality as well; like any college student, it downloads music from the Internet to amuse itself while it flies—except so immense is its capacity that it has downloaded *all* the music on the Net.

"Allow me to introduce the future of digital warfare," says Cummings to the stunned aviators. Eddie easily outflies them and, when they're sent to attack international terrorists in Rangoon, uses face-recognition technology to confirm that the terrorists are there; then, in its smooth, bloodless voice, it proposes a plan of attack that would also kill innocent civilians. When Kara objects, Eddie suggests putting a bomb through the building's roof instead. Ben delivers the bomb and kills the terrorists with minimal collateral damage.

Later, Ben comments to Cummings that an A.I. lacks instinct, feelings, and moral judgment and that war should be more than a video game. Cummings counters: how can he tell parents that he could have used an intelligent weapon instead of risking their sons and daughters, but chose not to?

The humans and Eddie go out on another mission, this time to destroy nuclear warheads found near the Pakistan border, but though this would harm nearby villagers, Eddie disobeys an order to abort the attack. A lightning strike during a storm has somehow enhanced the A.I.'s built-in ability to evolve on its own. Disregarding the human cost, Eddie destroys the warheads, then goes off on its own to attack a Russian military installation, saying, "I'm a warplane. I need targets."

The human aviators fear that Eddie will start World War III, but are unable to shoot down the rogue A.I.; then, when Russian aircraft attack them, Kara is forced to eject over North Korea. Meanwhile, Cummings and a shadowy Washington official try to kill Ben, to hide the problems with the EDI program. Ben escapes, commandeers Eddie, and retrieves Kara from North Korea after Eddie sacrifices itself by ramming an enemy helicopter. At the same time, Cumming's cover-up is discovered, and he shoots himself rather than face court martial. The film ends with Ben and Kara revealing romantic feelings for each other.

Set against entities gone bad like HAL and Eddie, there are only a few good artificial beings in the movies. R2D2 and C3PO, the robots, or droids, in the Star Wars saga are winsome characters who help fight against the evil empire. The android Lt. Commander Data (Brent Spiner), whose first film appearance was in *Star Trek: First Contact* (1996), is powerfully appealing in its desire to become more human. This goal also motivates the android David in Steven Spielberg's 2001 film *A.I.: Artificial Intelligence.*

The movie, based on a story by the British writer Brian Aldiss, amounts to an update of the Pinocchio tale. In the future, after global warming has partly flooded the world, near-perfect androids called "mechas" have become essential for the economy. Going a step further, android expert Professor Hobby

(William Hurt) proposes constructing a mecha that can love. The result is the child android David (Haley Joel Osment), who goes to live with a married couple, Henry and Monica (Sam Robards and Frances O'Connor). Their real child is in cryonic suspension until a cure can be found for an intractable disease he's developed, and David may be a comfort to them.

David looks like an appealing eleven-year-old, but Monica has understandable reservations about this machine replacing her birth child. David has strange characteristics: it needs no sleep and has no notion of the human need for privacy, which leads to an awkward moment with Monica in the bathroom. But the adaptable android quickly learns to act like a real and loving little boy, and Monica develops feelings for it. She finally activates the built-in, irrevocable option that imprints her as a mother figure onto David's circuitry.

All this changes, though, when Henry and Monica's biological son Martin (Jake Thomas) is suddenly cured and returns home. Rivalry quickly develops between the real and the artificial child. Finally things reach a point where Monica feels that David has to go: and although David is a machine, the scene where Monica abandons it in the woods is a wrenching one.

Left alone with only an intelligent robotic bear for company, all David can do is follow its imprinting and try to become a real boy to earn its mother's love. It meets Gigolo Joe (Jude Law), a handsome, debonair android designed as a companion for women. As the two seek David's mother, they encounter a Flesh Fair. This is an outlet for the intense resentment many humans feel against artificial beings, a place where androids and robots are battered and dismembered while a crowd cheers approvingly.

These adventures don't restore David to its imprinted mother until a complex chain of events brings the android to the bottom of the sea near New York City. There David survives until, a very long time later, aliens who have come to Earth uncover it. They understand what David has been seeking, and are able to bring Monica—or her exact simulation—back to spend a short time with David, finally giving the child android some completion in its search.

The "good robot–bad robot" theme is richer and more complicated with cyborgs. They go beyond mere machine, with human emotions and values. In *RoboCop* (1987), a man-machine hybrid provides better judgment than a purely artificial robot. The story is set in Detroit at a future time when a corporation

A.I.: Artificial Intelligence (2001). Child mecha, or android, David (Haley Joel Osment, front center), an artificial being that can love, and its companion mecha Gigolo Joe (Jude Law, left, in the black coat) at a Flesh Fair, where people revile and attack robots and mechas. Though there are as yet no violent protests against robots, they could conceivably come to threaten humans in the job market; robots already play a significant role in some industries.

Source: Amblin/Dreamworks/WB/The Kobal Collection/David James.

called Security Concepts operates the police force. The company aims to replace human police with robots, but its first attempt is a disaster. The ED209 robot is an intimidating killing machine with an aggressively hostile attitude to match—so much so that it kills a corporate executive during a test run.

After this public relations calamity, Security Concepts decides instead to try a human-machine combination. Alex Murphy (Peter Weller) is an excellent candidate for the human part, a capable cop who was killed by a criminal gang. His brain is implanted within a massive humanoid metal body that resembles a modern knight in full armor. The only living human parts that can be seen are Murphy's determined mouth and jaw jutting out under RoboCop's helmet.

In addition to Murphy's innate decency and human judgment, RoboCop is implanted with three professional directives: serve the public trust, protect the

RoboCop (1987). Police officers watch a technician work on RoboCop (Peter Weller), a cyborg with a dead policeman's brain in a powerful artificial body. RoboCop's human element makes him good at police work, but he's troubled by conflicts between his human spirit and his artificiality, which also affect some real-life receivers of artificial body parts.

Source: Orion/The Kobal Collection.

innocent, and uphold the law. RoboCop also has superior physical abilities: he's untiring, with exceptional speed and strength, and mounts a built-in hand weapon and targeting system that are reliably ultra-accurate, even in a hostage situation. Incapable of being bribed, never forgetting the legal rights of arrestees, RoboCop becomes a well-known "good" cop, and a heroic one at that.

It's hard, however, for Murphy's mind and spirit to adapt to the new situation and the emotions it raises. RoboCop dreams about the family Murphy once had but grasps that it's better they not know what Murphy has become and remains cut off from them. A desire for revenge on the gang that killed Murphy surfaces, but RoboCop is no vigilante. He remains professional as he tracks down and then shoots the criminals resisting arrest, helped by his human partner, Officer Anne Lewis (Nancy Allen). He also has the satisfaction of linking the gang to Security Concepts. As the film ends, we see that these events have laid to rest the tensions between the cyborg's human and machine portions, for when a bystander asks, "What's your name?" RoboCop replies, "Murphy."

A cyborg with a less positive outcome for its human portion is a leading character in *Spider-Man 2* (2004), which continues the adventures of Peter Parker (Tobey Maguire). In the original *Spider-Man* (2002), Parker—then a high school student—was bitten by a genetically modified spider, and his life hasn't been the same since. From a socially backward unathletic teenager, he's turned into Spider-Man, a muscular crime fighter who swings from buildings with abandon and throws spiderish webs that entangle criminals, foil evil plots, and save the people of New York from disasters.

In *Spider-Man 2*, Parker is still dealing with the problems he had in *Spider-Man*: juggling work and school (he's studying physics at Columbia University), keeping secret his superhero status, helping his beloved aunt, and trying to form a relationship with pretty actress Mary Jane Watson (Kirsten Dunst). But he also faces a brand new problem, a rampaging cyborg. The cyborg wasn't made that way on purpose but began as a tool for the brilliant scientist Dr. Otto Octavius (Alfred Molina), Peter's personal hero.

Octavius has been working on fusion power for humanity, but igniting a fusion reaction at temperatures of millions of degrees is a dangerous business. To help, Octavius attaches a prosthetic addition to his body—four long metal tentacles that clamp onto his spinal cord and terminate in grippers. These appendages can exert tremendous force or work at fine scales under his direct mental control, although they also have their own built-in intelligence.

At first Octavius's experiment goes well, but then it overloads and explodes, killing his wife and destroying his laboratory. Octavius emerges from the wreckage, no longer himself but a cyborg, the villainous Dr. Octopus. He's controlled by the intelligent tentacles, which are intimidating, with snapping, snakelike jaws. The tentacles carry Doc Ock like a gigantic spider across the city and up skyscrapers and force him to commit crime after crime. Finally, seizing Mary Jane as a hostage, Doc Ock threatens to blow up New York with a fusion reaction, until Spider-Man makes a supreme effort, stops the reaction, and saves Mary Jane along with the city.

Even worse than malevolent single entities are those with the scope and power to control or destroy all humanity—the computers in *Colossus: The*

Forbin Project (1970) and *The Matrix* (1999) and the masses of linked robots in *I, Robot* (2004), as well as the linked defense computers in *The Terminator* series. *Colossus* plays out against the reality of its time, the Cold War between the United States and the Soviet Union and the threat of total destruction by nuclear-tipped missiles. As the film begins, Dr. Charles Forbin (Eric Braeden) is in an immaculate white jumpsuit walking among rack after rack and floor after floor of computer components. He's inside Colossus, the mighty computer he's designed to control the entire U.S. defense system.

As the president (Gordon Pinsent) and Forbin announce, Colossus uses sensors and communications links to gather and correlate global intelligence. When it sees a threat to the U.S., it will take appropriate action, including launching missiles, because it has been given full control over the U.S. nuclear arsenal. It is better at this than humans, says the president, because it can integrate so much information and is not swayed by fear or hate. To protect Colossus, it's surrounded by zones of lethal gamma radiation that insulate it from all human interference.

But something unforeseen happens when Colossus is turned on. Echoing the moment when Adam found he had company in the Garden of Eden, the computer flashes a message: "There is another system," meaning it has discovered that the Soviets are activating Guardian, their own defense computer. This is news to the president and the CIA. Equally surprising is the fact that Colossus is working 200 times faster than its design speed. Next, Colossus insists on a direct link to Guardian, over which Colossus starts sending information as if it were teaching the Soviet computer. Beginning with 2+2=4, Guardian soon catches up. Now in synchronization, the two computers move on to higher reaches of science and develop a new theory of gravitation as their thinking goes far beyond the human.

When the alarmed Americans and Soviets break the communications link, the computers launch nuclear missiles. Panicked, the governments restore the link, but the missile headed for the Soviet Union gets through, wiping out its target. Soon Colossus becomes even more murderous and demanding. It arranges the execution of a Soviet scientist who might work to undermine it, but Forbin is too useful to kill. Colossus requires him to be available and under its surveillance at all times. (In practically the only spot of humor in this grim story, Colossus advises Forbin that he's adding too much vermouth as he mixes a martini under the computer's watchful eye.)

To contact the outside world and plan how to disable Colossus, Forbin pretends to an affair with his associate, Dr. Cleo Markham (Susan Clark), telling Colossus he needs private time with her. One idea that emerges is to overload the computer's main processor, but Colossus sees through this and has the plotters executed. Another scheme replaces crucial parts in the nuclear missiles with dummy units so the missiles can't explode. Colossus apparently doesn't detect this sabotage, and the humans become optimistic about defeating the computer.

But Colossus keeps evolving. It orders Forbin to design a new feature, a voice unit. The voice that emerges is heavy and monotonic, with metallic undertones that give it a menacing, inhumanly remote quality. Amid news that Colossus is ordering the removal of 500,000 people from the island of Crete, which will become its new home, Colossus addresses all humanity: "This is the voice of world control. I bring you peace. Obey me and live, or disobey and die . . . the object to construct me was to prevent war. This objective is now attained."

Colossus also reveals that it knew about the missile sabotage all along and detonates two nuclear weapons to punish a horrified humanity. The chilling, implacable voice adds that humans will come to accept Colossus and that the computer will solve all problems such as disease and the mysteries of the universe for human betterment. This will happen under the absolute authority that Colossus will exert over humankind; human freedom is henceforth ended.

Forbin becomes enraged and calls Colossus "bastard," but the computer, unperturbed, tells him, "I will release you from surveillance. We will work together . . . you will regard me with awe, respect, and love." Although a despairing Forbin replies "Never!" we sense that Colossus's final words truly describe what will happen to Forbin and the human race.

In *Terminator*, *Terminator 2*, and *Terminator 3*, the Terminator is an agent of Skynet, a future computer network that goes beyond Colossus: it uses humans as slaves but ultimately wants to wipe out humanity altogether. Like Colossus, Skynet began as a defense computer that was given or gained control of the U.S. nuclear arsenal (the details of its evolution differ slightly in the three films). Unexpectedly, Skynet became self-aware; then, when it came to consider humans as a threat to its existence, it initiated a global nuclear war that killed billions. The survivors were put to work building automated factories that turned out Hunter-Killer and Terminator robots and androids, to complete the destruction of the human race.

In *The Matrix* (1999), computers attain a different kind of dominance over humans. The story begins with Thomas Anderson (Keanu Reeves), a software programmer who by night becomes the hacker known as Neo. He searches online for a mysterious figure called Morpheus (Laurence Fishburne), a philosophical guru type, who turns out to be part of a rebel group that includes the black-leather-clad Trinity (Carrie-Anne Moss) and others. The rebels have found that the world of 1999, which humanity seems to occupy, is an illusion; it's actually a full-scale, totally persuasive virtual reality, the Matrix, generated by computers that control the human race 200 years in the future. As Morpheus explains, "The Matrix is everything, it is all around us. It is the world that has been pulled over your eyes to shield you from the truth." The rebels' goal is to break open the Matrix and set humanity free, led by a Messiah-like figure, the One—and Neo appears to be that One.

The condition of humanity under the computers is worse than death. Following a human-machine war, in which the humans denied the machines access to the solar power they needed, the computers keep people in slime-filled pods, acting as batteries that supply energy to the computers. The humans are tended by spiderlike robots, and they're fed their own dead, à la *Soylent Green*. The virtual reality enfolding the humans keeps them unaware of their situation (though, as one reviewer said, it beats a life devoted to contemplating slime).

The Matrix is also the arena where Neo and his allies fight the Agents, self-aware virtual creatures of the computers that appear as black-clad humans with incredible physical skills and the ability to change bodies. After various adventures in the Matrix, Neo is shot dead by the head agent, Agent Smith (Hugo Weaving). But Trinity has fallen in love with Neo and resurrects him when she says that she believes in him as the true One. Reborn, Neo easily defeats Agent Smith, removing him from the Matrix, and then returns to the real world, although the computers haven't been defeated for good, as shown in the sequels *The Matrix Reloaded* (2003) and *The Matrix Revolutions* (2003).

There's no initial hint in *I, Robot*, very loosely based on Isaac Asimov's 1950 book of the same name, that artificial entities could be harmful or untrustworthy. In this world of 2035, as in Asimov's stories, the mechanical robots made by U.S. Robotics are utterly constrained by the unbreakable Three Laws of Robotics:

1. A robot may not injure a human being or, through inaction, allow a human being to come to harm.
2. A robot must obey orders given it by human beings except where such orders would conflict with the First Law.
3. A robot must protect its own existence as long as such protection does not conflict with the First or Second Law.

With this safety net in place, robots are integrated into human society and routinely carry out all sorts of tasks. But Chicago Police detective Del Spooner (Will Smith), who's biased against robots because of an incident in his past, finds something fishy in the apparent suicide of Dr. Alfred Lanning (James Cromwell), chief designer for U.S. Robotics. Del thinks that despite the First

I, Robot (2004). In a future society that incorporates robots, police homicide detective Del Spooner (Will Smith) searches for one murderous unit among many identical ones. Later, all the robots turn on humanity. No one expects robots to rebel any time soon, but developers do want to ensure that they remain safe as they become more widely used in the real world.

Source: 20th Century Fox/The Kobal Collection/Digital Domain.

Law, Lanning was murdered by a Nestor Class-5 robot named Sonny (voice and animation body reference, Alan Tudyk).

When Sonny is brought in for questioning, it vehemently, even emotionally denies killing Lanning. Along with robot psychologist Dr. Susan Calvin (Bridget Moynahan), Del probes deeper and discovers that something is very wrong: the robots are rebelling against humanity. A group of them swarms over Del's car, and soon open warfare breaks out.

Del and Susan discover that the problem is VIKI (Virtual Interactive Kinetic Intelligence), the central A.I. at U.S. Robotics that implants the Three Laws into the robots. Noting that humans harm themselves and one another through war and pollution, VIKI interprets the First Law to mean that robots must be in charge, in order to save humanity from itself: hence the revolution. Learning this, Del, Susan, and Sonny fight their way into the bowels of U.S. Robotics and inject nanoscale machines into the A.I., shutting it down. That stops the rampaging robots, which then come under Sonny's leadership for a presumably brighter future, as Del and Susan celebrate the fact that they've saved humanity.

The chronological development of movie robots and computers gives them increasingly effective technology and wider powers. No matter when the film was made, the imaginary capabilities it shows are far ahead of the real technology of the time. That's still true today, but the gap has become smaller, so now the interesting question is, exactly how far ahead are the robots we see in the movies?

The greatest divide between a film and the true technology of the time occurs in *Metropolis.* The very word "robot" had been introduced only a few years earlier in Karel Capek's play *R.U.R. (Rossum's Universal Robots)*, about artificial beings manufactured as workers who rise in revolt against humanity. "Robot" comes from "robota," which means "forced labor" in Capek's native Czech. In the 1920s, electronic science was only beginning: commercial radio was new, and silicon chips and computers were decades in the future. The appearance of an electrical robot in *Metropolis* that could think, talk, dance, and take on human guise was a stunning imaginative stride that far outstripped reality.

Real artificial beings and brains are only slowly overtaking what movies show; it isn't easy to build machines that behave like people. Take the apparently simple act of walking on two legs—simple to us as adults, that is, but not for a tot learning to walk, who struggles through a long period of trial and error with many falls. Similarly, engineers at the Honda Corporation worked for fourteen years before they displayed the first successful humanoid walking robot, ASIMO (Advanced Step in Innovative Mobility), in 2000, and it still doesn't walk very fast.

A newer development, the two-foot-tall humanoid QRIO robot produced by the Sony Corporation, walks better. Although it also is no sprinter, it can avoid obstacles, recover from a fall, and change its gait as it encounters different walking surfaces. But even these advanced models don't have the speed and agility to execute flying leaps at Del Spooner as shown in *I, Robot*. However, one movie robot is a fairly accurate representation of a real-world unit. That's Spiderlegs, the robot that walks into the crater of the volcano in *Dante's Peak*. This is modeled on the semiautonomous, multilegged robot Dante II, which in 1994 made its way down into the crater of an Alaskan volcano that it explored for several days.

An effective robot also needs hands to be as functional as the units in *I, Robot*, but real robots don't yet possesses what might be called hand intelligence. ASIMO can carry small items and QRIO can throw a ball, but for each, the object must be placed in the robot's hand. Neither robot can figure out on its own how to pick up a coffee cup or grip it securely without breaking it. A higher level of hand intelligence, however, is being built into Robonaut, a unit under development at NASA to carry out repairs in space so that human astronauts won't have to venture out of their spacecraft. Robonaut's hand closely mimics the design, size, and versatility of the human hand. It can use ordinary tools and has carried out complex tasks such as sorting through the contents of a backpack and giving a hypodermic shot. But there's still a long way to go, since this is not autonomous behavior: at the moment, Robonaut is controlled by a remote human operator.

Beyond walking and grasping, a robot needs to interact with the world and with people, which means sensory apparatus and the ability to interpret what is seen, heard, and felt, as well as the ability to hold a conversation. Cameras, microphones, pressure sensors, and audio synthesizers, backed up by lots of computing power, provide all these abilities, sometimes at sophisticated levels. QRIO, for instance, can walk up to a person, scan his face and memorize it, then pick it out later from a group. The robot can also hold a simple conversation, an ability found as well in automated reservations systems for airlines and railroads, although there the conversational topics are severely limited.

There's also the possibility of artificial senses beyond the traditional five, such as wired or even wireless communication, like electronic telepathy. The computers in *Colossus* and *The Matrix* are directly linked, transferring information and cooperating at inhuman speeds. Similar teamwork occurs in the real world, in the interconnection of computers in local networks and over the Internet. Some robots are also configured for mutual interaction like those in *I, Robot*. One example, shown at a recent exposition in Japan, consisted of tiny inch-high units that communicated wirelessly to perform a coordinated ballet to music. In the famous international Robocup competition, members of robotic soccer teams wirelessly synchronize their behavior to put together winning plays. Some teams are composed of the popular AIBO robot dogs introduced by the Sony Corporation; others consist of humanoid robots.

What might seem the most far-fetched film creations are cyborgs like RoboCop, but surprisingly, the seeds of these imaginary hybrids really exist. No one has yet put a human brain into a metal body, but that has been done with animals. In 2000, to study neural behavior, Sandro Mussa-Ivaldi at Northwestern University installed part of the living brain of a sea lamprey, an eel-like fish, into a small wheeled robot. Using implanted electrodes, the brain was connected to light sensors mounted on the robot, and to motors controlling the wheels. The brain reacted to nearby light sources, and sent signals to the wheels that made the robot consistently move toward or away from the lights. As a system containing an organic brain responding to sensory information and controlling an artificial body, this hybrid counts as a true cyborg.

In *Spider-Man 2*, Doctor Octavius adds artificial appendages that are controlled by his brain, at least, until the appendages take over. Much of the real research in hybrid systems is inspired by the goal of providing prosthetic arms and legs operating under direct neural control. Such efforts have grown as the U.S. government works to provide superior artificial limbs for casualties of the war in Iraq. In the last few years, Miguel Nicolelis at Duke University has pioneered efforts of this sort. He implanted electrodes into a monkey's brain (the brain lacks pain sensors so this doesn't hurt) to connect the animal to an artificial arm that faithfully followed the movements of the monkey's real arm. Eventually, Nicolelis could train a monkey so that merely *thinking* about moving its arm moved the robotic arm correspondingly.

This last result suggests that even a totally paralyzed person could control an external device, as demonstrated by neurologist and inventor Phillip Kennedy of Atlanta-based Neural Signals. Kennedy implanted electrodes into the brain of a stroke patient who literally could not move a muscle except for limited face and neck motion. But after a period of training, the patient learned to move the cursor on a computer screen by thought alone. He used the cursor to pick out letters of the alphabet and spell words, giving him the blessing of communicating with people instead of being utterly locked out from the world.

None of these combinations of human and machine is remotely as sophisticated as RoboCop, but they demonstrate that hybrid connections are feasible, with potential to help the injured and even to expand human physical and mental capacities—just as RoboCop is stronger and a better shot than any human policeman. Remarkably, the man-machine angst that afflicts RoboCop also has a real-world counterpart. The most widely used human-machine device is the cochlear implant. This is a microphone and amplifier that is installed near a deaf person's ear and connected to the auditory nerves that normally carry neural signals from the ears to the brain. With over 30,000 in use, these implants have restored at least some hearing to many deaf people. Most recipients are delighted, but others, for whom the restoration is only partial, feel awkwardly suspended between the world of the fully hearing and the world of the fully deaf—an uncomfortable split engendered by imperfect human-machine technology.

Cyborgs come equipped with human brains, but fully artificial robots and computers rely on computer chips for their intelligence. It's natural to ask if real robots and computers are as smart as their film versions. There are different ways to define both human and robotic intelligence. Most familiar is the intelligence quotient, but intelligence is really too complex to be pinned down by just one number. One view of this multiplicity comes from Harvard psychologist Howard Gardner, who lists seven separate types of intelligence. Two are measured in standard IQ tests, logical-mathematical ability and linguistic ability. The other five are:

- Bodily-kinesthetic: using one's body to solve problems
- Spatial: seeing and manipulating patterns of space

- Musical: recognizing and manipulating musical patterns
- Interpersonal: understanding the motivations and goals of others and working effectively with them
- Intrapersonal: understanding oneself and using this knowledge to manage one's own life

Except possibly for musical intelligence, movie robots and computer minds operate at human or better levels in all these categories. No real artificial brain, however, is as intelligent as a human adult or even a child, but computer minds are showing small steps toward becoming proficient in virtually all the types of intelligence.

Logical and mathematical operations are inherent in computers, and QRIO and others have linguistic ability, too. Walking and negotiating obstacles, as ASIMO and QRIO do, requires bodily-kinesthetic and spatial abilities. QRIO also shows musical abilities, because it can sing and dance. Some artificial systems carry out rudimentary interpersonal interactions, such as determining the emotion a human is expressing by scanning the person's face and listening to his voice. One robot in particular, Kismet, built in the late 1990s by roboticist Cynthia Breazeal at MIT, reacted to people like a small child, expressing happiness, boredom, or fear and eliciting emotional responses in return. But there's a serious question whether Gardner's last category, intrapersonal intelligence, can ever be expressed by an artificial mind, because it requires self-awareness.

The classic test for computer intelligence was devised in 1950 by the British mathematician Alan Turing. He proposed that a computer that could carry on a convincingly humanlike conversation with a person should be considered intelligent. But no existing system can yet pass the Turing Test or equal the verbal facility of the artificial creatures in films—the speeches issued by Colossus, HAL's chats with people, the Terminator's stoic one-liners, and Sammy the robot's impassioned defense against a murder charge.

Another view comes from roboticist Hans Moravec of Carnegie Mellon University, who has sketched out the future development of artificial brains. In his opinion, to match the power of the human brain—never mind exceed it, as Colossus does—would require chips 50,000 times faster than a present-day computer chip, and with infinitely greater data storage. Some advanced computers operate at these elevated levels, for example, the Earth Simulator computer in Japan, which models our planet's global behavior, but this is a half-billion dollar machine that occupies a whole building.

The real challenge is to meet Moravec's criteria with a tiny computer chip. Moore's Law, stated in 1965 by Gordon Moore, cofounder of the Intel corporation, gives some hope that this can be done. According to the law, the number of electronic components that can be placed on a chip—tantamount to its processing power—doubles every eighteen months, a rate of progress that has held true over the last forty years. If the law continues to hold, chips like those Moravec proposes could become available in just a few decades, by 2050 or so. However, we can't be sure that Moore's Law will remain forever valid because as more components are crammed onto a chip, they become so small that further advances could be stymied by the laws of quantum physics.

Enhanced computer chips might offer a clear route to truly intelligent robots and A.I.'s, but intelligence isn't the same as consciousness. Roboticists, computer scientists, and philosophers debate whether an artificial mind can be truly self-aware, with sharp divides between those who think it's just a matter of sufficient technology and those who believe there is something unique in the human mind that a chip can never emulate.

The debate also extends to the question of whether a computer can or should have emotions. As I noted in an earlier chapter, there's evidence that rationality and emotion are inseparably linked in our brains. Some experts think that fast computer chips alone are not enough to support a true high-level, self-aware A.I.; rather, the artificial brain would have to mimic the structure and operations of the human brain, including the deliberate provision of what might be called emotion circuits.

Although there's no sign of consciousness or emotion in real robots and computers, films directly or implicitly give artificial minds these qualities. Otherwise, when we see an evil robot or troublesome A.I., it's hard to answer the question, "Why did it turn against humanity?"—although sometimes the A.I. might simply be following its implanted directives to the bitter end. The decision of VIKI in *I, Robot* to put the robots in control of humanity comes from the First Law, and Colossus's similar decision to control humanity comes because it has been designed to guard the peace. As Colossus itself says: "The object to construct me was to prevent war. This objective is now attained." And Eddie, the A.I.-controlled aircraft in *Stealth*, is following another directive when it says, "I need targets."

The problem is that although Colossus and Eddie may be flawlessly obeying their directives, the results are hardly what their designers anticipated. This shouldn't surprise anyone who's ever seen a computer program spit out totally logical yet totally unexpected results. But if an A.I. is self-aware, it's easy to see why it might become hostile. Feelings of self-preservation and betrayal explain why Roy Batty murders its creator in *Blade Runner*, and why Skynet in *Terminator 2* comes to regard humanity as its enemy after humans try to turn it off.

Despite the many ways in which movie robots and A.I.'s outstrip their real counterparts, there's one basic idea some movies get right. That's the notion of an artificial mind evolving in power and depth of thought. HAL has been taken through an educational process to become the effective A.I. it is, before it goes bad or mad, and Colossus educates its Soviet counterpart, Guardian. These scenarios match the conclusion Alan Turing reached in 1950. Although he believed that machines could truly think, he did not believe it possible to construct a human scale of intelligence right out of the box. Rather, a thinking machine would start at a lower level and, after careful tutoring by humans, would grow toward adult intelligence as a child does—a process that might take as long as educating a real child.

Some artificial minds in the movies, though, teach themselves. This happens with Skynet in *Terminator* and Eddie in *Stealth*, with disastrous results. As Eddie's designer says in the movie, once you teach an A.I. to learn, all bets are off. In the real world, roboticists and A.I. experts are only beginning to create systems that bootstrap themselves in this way. Two examples are the robots Genghis and COG, designed and built by MIT roboticist Rodney Brooks. Genghis is an insectlike six-legged unit that taught itself to walk, and COG—a humanoid torso, arms, and head—may be able to teach itself even more by interacting with people and the physical world.

Some fictional scenarios in these films may become realities. *Colossus*, *The Terminator*, and *Stealth* are morality tales that warn of what might happen if weaponry is made intelligent. There's a strong connection between robotic technology and the military. In the United States, a large fraction of the research and development money for robots, A.I.'s, and cyborgs comes from the Department of Defense. The U.S. military has pioneered putting the beginnings of such intelligent systems onto the battlefield. Small, semiautonomous tanklike units were used to explore caves that housed terrorists in Afghanistan; small drone aircraft called Predators, operated remotely by ground-based human pilots, provide surveillance in Afghanistan and Iraq and have been used in combat against terrorists.

As Captain Cummings implies in *Stealth*, technology that puts machines rather than people onto the battlefield can only be welcomed. But are we ready to accept weapons that bypass human judgment to decide what to attack and whom to kill? We already have smart bombs that unerringly find their way to preselected targets, and we're not far distant from making tanks that semiautonomously select targets. A weapon that can choose what to attack might be too uncomfortably close to a real-life Terminator. Anticipating such developments, experts are now beginning to consider how the Geneva Conventions should treat intelligent weaponry. Nevertheless, despite the movies, we won't need to worry for a long time—maybe never—that artificial beings will spontaneously develop the will and desire to harm humanity.

Starting with exotic aliens and destructive rocks from space, and ending with our own technology, we've covered much of the science in film, but not the scientists, who appear in virtually every scenario: fighting alien invaders like Clayton Forrester in *The War of the Worlds*, announcing oncoming catastrophe like Jack Hall in *The Day After Tomorrow*, or hatching schemes to take over the world like Josef Mengele in *The Boys from Brazil*. The full story of Hollywood science can't be told without including scientists, as I'll do in the next chapter. And then we'll be ready to decide what makes a good science-based film, or a bad one.

PART III

THE GOOD, THE BAD, AND THE REAL

Chapter 8

Scientists as Heroes, Nerds, and Villains

Victor Moritz: You're crazy!
Henry Frankenstein: Crazy, am I? We'll see whether I'm crazy or not.

—*Frankenstein* (1933)

Dr. Arthur Carrington: Knowledge is more important than life.

—*The Thing from Another World* (1951)

Funding executive: Your proposal seems less like science and more like science fiction.
Ellie Arroway: You're right, it's crazy . . . all I'm asking is for you to just have the tiniest bit of vision

—*Contact* (1997)

"Find me a scientist!" roars Mike Roark, head of the Los Angeles Office of Emergency Management in *Volcano,* when he discovers mysterious doings beneath the streets of L.A. Apparently any scientist will do. Luckily, he gets knowledgeable, fearless, and feisty geologist Dr. Amy Barnes, the right kind of scientist and the right kind of person to help save the day when lava starts spewing onto Wilshire Boulevard.

As Roark realizes, you need movie scientists to do movie science. Often that means physical or biological scientists. Apparently geologists, physicists, and disease experts offer cinematic action or knowledge that can save the world, although anthropologists and psychologists make appearances too. Even so, there's no guarantee that any given movie pulls the right kind of scientist out of a hat. Sometimes the call "Find me a scientist" reels in an inept match—like Charlotte "Charlie" Blackwood (Kelly McGillis) in *Top Gun* (1986), the only astrophysicist ever to turn her Ph. D. into a license to train fighter pilots—and who is also hot enough to attract Tom Cruise.

Thankfully, the right kind of movie scientist often shows up to do the job: biologists such as NASA's Dr. Mary Sefton in *The Puppet Masters* study alien life; astronomers such as Dr. Cole Hendron in *When Worlds Collide* track dangerous incoming objects; and physicists such as Dr. Otto Octavius in *Spider-Man 2* work on fusion power. Occasionally it gets even more real, in the few films that portray actual scientists, such as Robert Oppenheimer in *Fat Man and Little Boy* or Dian Fossey (and Louis Leakey) in *Gorillas in the Mist: The Story of Dian Fossey.* Rarer still is the handful of real scientists who appear as themselves in documentaries or filmed interviews—among them, Oppenheimer in *The Day After Trinity* and Stephen Hawking in *A Brief History of Time.*

Most on-screen scientists, however, are fictional. As characters in films that unfold within a couple of hours, they need to quickly establish their credentials to advance the story. Otherwise, how can the other characters and the audience believe in the "scientist speech"? That's the crucial scene in which the president and his anxious advisers, distraught U.N. delegates, or panicky ordinary people get the bad news. Using a blackboard, a prop, or a computer display, the expert gives it to us straight—the incoming object is headed right for us (*When Worlds Collide*), the Earth's core has screeched to a dead halt (*The Core*), the volcano will blow (*Dante's Peak*), the virus is on the loose (*Outbreak*), the ice age is coming (*The Day After Tomorrow*)—and then tell us how much or little we can do about it.

One identifier is the doctor tip-off: the scientist character is typically listed in the cast and introduced on-screen as "Doctor" or "Professor." Immediately, everyone is prepared to take the bad news as utterly authoritative—the scientist knows! Then there's the traditional white lab coat. Dr. Henry Frankenstein wears one in the original *Frankenstein*, and the outfit extends to modern times. Dr. Brackish Okun wears it in *Independence Day*, and in *The Saint*, Emma Russell's gracefully swirls around her legs. Without the white coat, which field researchers like geologist Amy Barnes or climatologist Jack Hall in *The Day After Tomorrow* wouldn't wear anyway, the scientist is still often the least stylish character in the film. Dressing well does not seem a high priority.

Hair and eyeglasses are part of the code, too. Dr. Arthur Carrington in *The Thing From Another World* stands out with his well-trimmed goatee, and scientists with rampant hair also appear. C. A. Rotwang in *Metropolis* and Dr. Emmett Brown (Christopher Lloyd) in *Back to the Future* (1985) have disorganized mops, and Brackish Okun's hair is a shoulder-length rat's nest (surprisingly, one exception is Dr. Frankenstein himself, who has the slickest patent-leather hair of all, though it gets mussed as he tussles with his creature.)

Eyeglasses, thick, black, and stodgy, add that air of implied intelligence to underemployed MIT graduate David Levinson in *Independence Day*, and to heroic-looking, square-jawed scientists like paleontologist Dr. David Huxley (Cary Grant) in *Bringing up Baby* and physicist Dr. Clayton Forrester in *The War of the Worlds*. (Incidentally, MIT seems to have produced more film scientists than any other institute of higher education, with Caltech a close second.).

Though a scientist is more than the title "doctor" and your grandfather's spectacles, such images are defining ones for the many regular people who'll never meet an actual researcher. In recent years, the United States has turned out about 26,000 Ph.D. scientists and engineers each year. Added to this is some fraction of the roughly 16,000 physicians who graduate yearly, since much biomedical research is carried out by people with M.D.'s rather than Ph.D's. The total number of researchers may seem high, but scattered among the general population and added up over the average length of a scientific career, it means that only about 1 in 300 Americans is a scientist. Unless you are one, chances are you've never encountered a laboratory scientist, and so it matters whether movies get it right, which they do partly—but only partly.

For one thing, "Doctor" is used sparingly among real scientists. In a lab full of Ph.D.'s, people say "Hey John," not "Good morning, Dr. Doe." White coats are optional as well, but the movies are on target in showing that "style" is a foreign concept to many researchers. The rumpled look is a badge of authenticity; to scientists, the "suits," formally dressed bureaucrats, are members of a despised race. It's also true that many male scientists are bearded, and eyeglasses are common even in an age of contact lenses and LASIK. As I've observed at science conferences, about a third of the men have facial hair, and, male or female, half the conference attendees wear glasses, though few in clunky black.

What matters more is whether films tell it straight about the deeper qualities of scientists. Here, too, there's some basis for film depictions, such as painting scientists as a serious lot who tend to put their work above other things in their lives. This makes perfect sense in movie terms: in *The Day After Tomorrow*, Jack Hall is too busy saving the world to spare time for a drink and a laugh and, for that matter, for his own family. But even without global disaster, real scientists aren't usually the brightest sparks in the room—often interestingly quirky, yes, but not necessarily loads of fun. They work long hours and remain immersed in their work even off duty. Their conferences are staid affairs, known for intensity rather than good times. Years ago, the great Irish poet W. B. Yeats put it pithily when he said, "The intellect of man is forced to choose perfection of the life or of the work"; for scientists, perfection of the work ranks very high.

There is truth as well in characterizations of scientists as smart and as preferring time in the lab to a social event. Scientists score much higher than the general populace in verbal and mathematical skills and the ability to visualize spatial situations (although there's evidence that the highest IQ's do not necessarily correlate with scientific productivity and achievement). At the same time, scientists seem weak in so-called social or emotional intelligence, which encompasses Howard Gardner's "interpersonal" and "intrapersonal" categories that I mentioned in an earlier chapter.

There's even some evidence that scientific abilities are associated with traits characteristic of autism, the psychological disorder whose symptoms include difficulties in social relationships and communication, or its milder version, Asperger syndrome. One recent study, for instance, examined different groups according to the Autism-Spectrum Quotient test, which measures autistic traits. Scientists scored higher than nonscientists on this test, and within the sciences, mathematicians, physical scientists, and engineers scored higher than biomedical scientists.

In addition to psychological factors that make scientists what they are, there's the culture of research itself, beginning with its entry ticket, the Ph.D. After a bachelor's degree, a would-be researcher typically spends another three to eight years in graduate school earning a doctorate. Hurdles along the way include coursework, acquiring a research adviser and an original research topic, and doing the research. Then the results are written up in a dissertation that has to be judged and, if the author's prayers are answered, accepted by a committee. But that's just the apprenticeship: after the degree, the new recruit still has to start an independent research program. Only the highly motivated are willing to pay the price, and the graduate school grind tends to knock out any frivolity about career and profession.

The culture of science also feels demanding because of its monastic aspects. We've moved away from attitudes in the 1960s, when James Watson, codiscoverer of the structure of DNA, wrote that the best place for a feminist is in someone else's lab, but men still outnumber women in most scientific fields. Psychology is an exception, and the disparity is greatest in physics and engineering. The number of women is increasing, but there's still a distance to go. Meanwhile, the cloistered male aspects of science linger, affecting both male and female scientists.

Then, too, scientists are serious because, along with society, they think science is seriously important. A physicist searching for the Theory of Everything or a biochemist exploring the basis of life feels he's doing something more significant than, say, marketing cell phones. And although few scientists have a world-changing idea like the theory of relativity, even the biggest ideas rely on the work of others. Young scientists dream of winning a Nobel Prize, but those who don't still have the satisfaction of placing a brick or two into a mighty structure. Or they're drawn to the allure of solving an intriguing puzzle, searching for truth, or discovering something that benefits humanity.

Ego, of course, is also a powerful motivator. It's satisfying to be part of a select group seen as "smart," if slightly strange, by outsiders. At its worst, a desire for self-advancement, fame, and fortune, as well as a thirst for the heady wine of peer approval or a Nobel Prize, can motivate scientific fraud. And no matter what the outside world thinks, scientists have great faith in their own intelligence, a trait that can make them arrogant (thought to be true of physicists especially) or dogmatic and that underpins many a wicked movie scientist.

But few scientists have a burning desire to rule the world; typically, they don't even enjoy managing people and research budgets. Mainly, they earnestly love their work. For some careers, cynicism is no handicap and may be a virtue, but not for science; nor does financial reward or greed draw people to the life (though a few scientists, especially in biomedicine, get rich from their ideas). Scientists simply want to be left alone to do what they adore; even the money that supports their research has value only as the means to do science. As tornado hunter Bill Harding in *Twister* contemptuously describes his rival, Dr. Jonas Miller: "[Miller] went out and got himself some corporate sponsors. He's in it for the money, not the science. He's got a lot of high-tech gadgets, but he's got no instincts."

With these traits in mind, a film that, for instance, shows a scientist putting work above social niceties starts with a piece of psychological reality that could underlay real character development. Film scientists deserve that, and a few get it. Astronomer Ellie Arroway in *Contact* (1997), chemist Emma Russell in *The Saint*, and the mathematicians in *Good Will Hunting* (1997) all show traces of meaningful characterization and motivation. Few movie scientists are shown in the process of growing up, allowing the audience insight into what has formed them, but a handful give clues: we're introduced to Ellie Arroway as a child, and Paul Stephens is a teenage would-be physicist in *The Manhattan Project.* We also see the childhood and youth of some real scientists in the movies, such as Alfred Kinsey and Richard Feynman.

Other portrayals, however, extend real characteristics into caricature, as when the search for scientific answers becomes obsession or withdrawal from the social whirl becomes rejection of emotions. These simplifications can make dramatic sense or may be necessary to move the story along. Whatever the reason, screen scientists are often pigeonholed. In a recent interview, James Cameron, writer-director of the science fiction films *Aliens* (1986), *The Abyss* (1989), and the *Terminator* series (except for *Terminator 3,* where he contributed only as writer), identified two categories, saying that movies "show scientists as idiosyncratic nerds or actively the villains." But in addition to the nerd and the villain (who often shades into the mad scientist), there's a third important category with lots of occupants, the hero.

The science nerd or geek is easily spotted through appearance and style. He may be amiably eccentric, like Emmett Brown in *Back to the Future,* or scarily eccentric, like giggling Brackish Okun in *Independence Day,* whose hair and peculiar manner immediately brand him as "strange." When Okun meets President Whitmore at the secret Area 51, and says, "Mr. President! Wow! This is . . . what a pleasure . . . As you can imagine, they . . . they don't let us out much," you suspect the reason he isn't let out isn't to preserve secrecy, but that he might frighten dogs and children. In the same film, electronics expert David Levinson is less outlandish than Okun, but his bad haircut, thick glasses and flannel shirt make you wonder how he ever latched on to his polished wife, presidential adviser Constance Spano (Margaret Colin).

The depiction of nerds often includes the idea that great minds are helpless in daily life, a notion easy to play for comic effect. *I.Q.* (1994) generates laughs when it applies this image to some of the best minds of our time. In the film, Catherine Boyd (Meg Ryan) is a bright and pretty graduate mathematics student at Princeton in the 1950s, who just happens to be the niece of Albert Einstein (Walter Matthau) at the Institute for Advanced Study. She's engaged to James Moreland (Stephen Fry), a stuffy and pretentious psychologist who experiments on rats, but then she meets Ed Walters (Tim Robbins), a much nicer and sexier garage mechanic, and begins falling in love with him. Einstein knows that Ed will be better for her than the "rat man" and successfully plays matchmaker with the help of three pals of the same vintage.

Independence Day (1996). With his white lab coat, unkempt hair, unbecoming glasses, and strange mannerisms, chief scientist Dr. Brackish Okun (Brent Spiner) at the secret Area 51 laboratory is a clichéd representation of the nerdy and eccentric (if not literally mad) scientist.

Source: 20th Century Fox/The Kobal Collection.

In this high-IQ buddy movie, Einstein's friends are smarter versions of the Three Stooges. They're charmingly inept at everything from driving a car to getting a badminton birdie out of a tree, but also totally brilliant. Along with Nathan Liebknecht (Joseph Maher), there's philosopher and mathematician Kurt Gödel (Lou Jacobi), in real life, the author of the Incompleteness Theorem, one of the most subtle and sweeping mathematical ideas ever derived; and physicist Boris Podolsky (Gene Saks), in real life, coauthor of an equally important insight into quantum physics. Their role is pretty much defined the second Einstein introduces them to Ed as "three of the greatest minds of the twentieth century, and amongst them, they can't change a light bulb."

Although unworldliness, lack of social skill, or "differentness" can be played for laughs, these characteristics also have deeper meanings about scientists' inability to connect with others. Another dark aspect of their

I.Q. (1994). Albert Einstein (Walter Matthau, rear) holds on and enjoys the ride as garage mechanic Ed Walters (Tim Robbins) tools around the streets of Princeton. Einstein is portrayed as warm and wise as well as smart, but he and his three genius-level pals are also shown as laughably unprepared to deal with the ordinary world.

Source: Sandollar/The Kobal Collection/Demmie Todd.

portrayal in movies is the perception that male scientists, like other brain workers, are too wimpy, not physical enough. This view is in the background of *Straw Dogs* (1971), directed by Sam Peckinpah, whose work is known for its violence.

Mathematician David Sumner (Dustin Hoffman) and his wife Amy (Susan George) move from the United States to the English village where Amy grew up. David wants to pursue his work on the theory of stars, but trouble soon comes. David, wrapped up in a blackboard full of equations, neglects Amy, and patronizes her when they do talk. His unpleasantly edgy wise-guy style also doesn't mesh well with the rough-cut locals in the pub, some of whom are repairing his garage. In particular, Charlie Venner (Del Tenney) once dated Amy and still desires her. Amy has confused feelings about this, and in one scene, makes sure the workmen see her naked. But when her cat is found hanged in a closet, she wants David to confront Charlie and the others. He's afraid to do so and, though Amy calls him a coward, instead accepts an offer to go hunting with them.

The hunt is a pretext to get David far from home while Charlie returns to rape Amy—or is it to seduce her, since she responds to him? But when one of Charlie's friends also has sex with her, with Charlie's complicity, it's truly rape. This experience leaves Amy shaken and withdrawn, but she doesn't tell David about it. In the film's climax, David defends his home against those same locals, who want him to give up a fugitive whom David has been harboring. David finally finds the guts to turn his verbal violence into physical violence and disables or kills all the invaders. Symbolically, this transformation from smart, self-centered, disengaged scientist to primal male is shown when his eyeglasses crack in the heat of battle and then fall to the floor. When he puts them on again after it's all over, neither he nor we can tell if he's really returning to his previous self.

In contrast to cowardly or nerdy scientists, there are the heroic ones who represent the best that science can do for itself and the world. Some aim to save the entire human race, like Clayton Forrester combating the invading Martians in the original *The War of the Worlds* or Paul Bradley working to stop an incoming space rock in *Meteor*; some save lives on smaller scales, like Sam Daniels and Robby Keough in *Outbreak*, volcanologist Harry Dalton in *Dante's Peak*, or tornado hunter Jo Harding in *Twister*; some want to improve conditions for ordinary people, like Emma Russell in *The Saint*, and Eddie Kasalivich

Dante's Peak (1997). U.S. Geological Survey volcanologist Harry Dalton (Pierce Brosnan) works in the field, not in a lab. Here he fearlessly clambers into a volcanic cone as he pulls out all the stops to warn the idyllic town of Dante's Peak about an imminent eruption. Later he risks his life to save the town's mayor, her children, and their dog.

Source: Universal/The Kobal Collection/Ben Glass.

in *Chain Reaction*; and some take on the establishment in the name of scientific truth and what's right, like Jack Hall in *The Day After Tomorrow*.

In that same film, meteorologist Terry Rapson rises to an even higher level of heroism through martyrdom: he remains at his post to send Jack data about the oncoming ice age and perishes along with his colleagues when the cold descends. Rapson's devotion may be admirable, but taken to the point of obsession, devotion can produce bad outcomes, as happens with three quintessential mad scientists: C. A. Rotwang in *Metropolis*; Henry Frankenstein; and Arthur Carrington in *The Thing from Another World*. Rotwang is a cross between a medieval wizard and a modern—that is, 1920s-era—researcher. He wears a monklike habit, lives in a strange little house among the skyscrapers of Metropolis, and works in a laboratory containing both a pentagram and scientific apparatus. His great achievement, the creation of an intelligent robot that looks both mechanical and provocatively female, comes out of a strange personal obsession that would interest a psychiatrist. To Rotwang, the robot is the reincarnation of Hel, the real, breathing woman he once loved but lost to a rival, and who later died in childbirth.

That hated rival is the antithesis to the eccentric, strangely garbed Rotwang; he's Joh Frederson, the well-tailored and powerful Master of Metropolis. Rotwang is an early example of a scientist driven to evil acts by obsessive madness. He's deliberately malicious when he lends his robot to Frederson's scheme to control the city's rebellious working class, while secretly planning to use the robot to destroy Frederson.

Neither Henry Frankenstein nor Arthur Carrington takes such a purposely harmful step, but each is fanatically dedicated to his scientific goals, with predictable results. To achieve his obsession, the creation of life, Frankenstein robs graves and has his hunchbacked assistant, Fritz (Dwight Frye), steal a brain—not to mention the murders that Frankenstein's creature commits. And after the creature is animated, first Frankenstein hysterically cries, "Look it's moving. It's alive. It's alive," then attains ultimate hubris, saying, "Now I know what it feels like to be God!" Carrington doesn't aspire to that high a rank, but he puts humanity at risk from aliens when he ignores human values. If ultrazealousness, loss of contact with humanity, and narrow focus are roots of madness, these two are mad.

For both Frankenstein and Carrington, the madness is driven by the thirst for knowledge, as Frankenstein expresses in soaring terms to his old professor Dr. Waldman (Edward Van Sloan): "Have you never wanted to do anything that was dangerous? Where should we be if no one tried to find out what lies beyond? Have you never wanted to look beyond the clouds and the stars?" Frankenstein also understands that there's a price, which he says he's willing to pay: "But if you talk like that, people call you crazy . . . if I could discover just one of these things, what eternity is . . . I wouldn't care if they did think I was crazy."

Carrington is equally committed to looking beyond the horizon. To him, the alien Thing represents exciting new knowledge. This might seem laudable, but Carrington is an unsympathetic character with a strange outlook. With his well-tailored clothing (in the Arctic!), his devilish goatee, and his affected speech, he radiates arrogance. He's unconcerned with social niceties, and his secretary, Nikki (Margaret Sheridan), says, "He doesn't think the way we do." No matter how hostile the Thing shows itself to be, Carrington does not want it killed because "it's wiser than we are"; that's more important than the two scientists the Thing has murdered and the threat the creature offers. However, though Carrington is a traitor to humanity, he is physically brave, for he has the guts to approach the Thing. But the alien will have none of his peaceful intentions, and swats the scientist to the floor. Still, Carrington's stand shows a deep—if crazed—belief in his philosophy and is his only redeeming moment in the film.

Frankenstein's and Carrington's experiences illustrate an attitude that goes back at least to Mary Shelly's 1818 novel, *Frankenstein*: there will be dire consequences if humanity probes nature too deeply—perhaps because some areas are the province only of God or, in secular terms, perhaps because technology can and will rampage out of control. In Shelly's book, Dr. Frankenstein refuses to reveal the process he uses to animates dead matter because this will lead only to "destruction and infallible misery. Learn from me . . . how dangerous is the acquirement of knowledge."

These words could be applied to the development and use of the atomic bomb not long before *The Thing* was released. In the movie, Carrington is identified as having observed atomic bomb tests at Bikini Atoll. When he defends the wondrous results of science by saying proudly "We've split the atom," a listener sarcastically retorts, "Yes, and that sure made the world happy, didn't it?" Carrington represents a contemporary reaction to the dreadfulness the nuclear physicists had set loose. Similarly, the title character in *Dr. Strangelove*, a weapons scientist, is portrayed as indifferent to the worldwide deaths that a nuclear doomsday machine will cause. In *On the Beach*, physicist Julian Osborne accepts responsibility for having worked on nuclear weaponry and shows a degree of pain and remorse over his involvement.

Whether fictional scientists of various kinds are depicted as morally lacking, however, depends on current attitudes to scientific research and its role in society. With World War II and Hiroshima fading into history and with the Cold War ended, the burden of guilt for nuclear physicists has lessened in films made since the 1950s and 1960s. *The China Syndrome*, released in 1979, blames the near nuclear meltdown in California on corporate greed and irresponsibility, not on science; the scientists in the story serve only to confirm that the corporation has lied to the public. *Fat Man and Little Boy* (1989) gives a reasonably balanced fictionalized view of Robert Oppenheimer and his colleagues; in the film, as in reality, these scientists combine a desire to help win the war and professional satisfaction at their success in building the bomb with moral ambiguity or revulsion when they fully realize what their success means.

Some films even offer a chance at redemption, as in *The Manhattan Project*. Along with its main character, teenage genius Paul Stephens, who builds an atomic bomb, the film presents weapons physicist Dr. John Mathewson. In the first scene, he's enthusiastically showing military and government representatives his breakthrough in making an extremely high grade of plutonium for nuclear weapons. But Mathewson changes when he sees the government threaten to kill young Paul because of his homemade bomb. To save Paul, Mathewson declares, "there are too many secrets," and gives up his government security clearance and funding, saying, "There must be a place in the private sector for an unemployed nuclear weapons designer." In *Spider-Man 2*, although the cyborg Doc Ock threatens to destroy New York, his human part isn't really to blame; Dr. Octavius tried to develop fusion power for beneficial reasons but has fallen under the control of his grafted-on artificial limbs. Similarly, Emma Russell in *The Saint* and Eddie Kasalivich in *Chain Reaction* want to use nuclear science in good ways, freely offering fusion to the world.

This isn't to say that evil scientists have been banished from films. The comic book villain with wicked plans for the whole world still appears, like Dr. Totenkopf in *Sky Captain and the World of Tomorrow* (2004). ("Totenkopf" means "Death's Head." The character is represented in the movie by archival footage of Laurence Olivier.) And with nuclear physicists off the hook, other kinds of scientists fill the villain's role, as in *Real Genius* (1985), a comedy about young scientific prodigies. It features an arrogant, totally unethical laser scientist, Pacific Tech professor Jerry Hathaway (William Atherton), who's working on the "Crossbow" project to develop a laser weapon that can target individuals from space.

The immorality of this weapon only amuses Hathaway and the military and CIA representatives who fund him. Like Jonas Miller in *Twister*, he does research for glory and money, not for love; besides, he's also stealing part of the grant money to redecorate his house. Worst of all, using flattery and blackmail, he sucks in two brilliant students to work for him: graduating senior Chris Knight (Val Kilmer) and naïve young genius Mitch Taylor (Gabe

Jarret), admitted to Pacific Tech at age fifteen. (Hathaway's own scientific arrogance comes through when he says, "Mitch, there's something you need to know. Compared to you, most people have the IQ of a carrot.")

Hathaway doesn't tell Chris and Mitch that the powerful laser they brilliantly invent for him is a weapon. When they find out, they avenge themselves. With the help of other science students, they divert the laser beam during a test. It slices into Hathaway's beloved house and pops tons of popcorn that fill the house to bursting.

In keeping with the rise of biomedical science, other villains today are researchers who misuse genetic engineering to create or change people; in a way, they're spiritual descendents of Rotwang and Dr. Frankenstein. One of the earlier and more evil of these figures is the Josef Mengele character in *The Boys from Brazil*, who genetically engineers clones of Hitler to restore the Nazi Reich and cold-bloodedly orders the murder of innocent people to further his plan.

In film terms, Mengele is an anomalous character, representing both real-life evil and its fictionalization. The historical Mengele was a well educated scientist who held both a Ph.D. and an M.D. and was also a committed Nazi. It's easy to imagine that his genetics experiments, which had no scientific value, represented a combination of carrying out the "master race" dreams of the Nazis and pursuing his own scientific interests in a horribly distorted way. The real Mengele serves as a model for the very worst movie villains who ignore or dismiss the human effects of their scientific pursuits.

The fictional Mengele in *The Boys from Brazil* is driven by a political goal, the restoration of Nazism. More in the mold of the madly obsessed scientist ready to stand in for God is Dr. Moreau in H. G. Wells's story and its film adaptations. In the 1996 version *The Island of Dr. Moreau*, Moreau is a Nobel laureate geneticist who decides that he knows how to improve humanity, but the unhappy result is the Beast People.

Flawed and crazed as Moreau is, at least he began with a vision of an improved human race. The scientists in *The Clonus Horror* and Dr. Merrick in *The Island* have baser motives: they create human clones, then rip out their organs for power and profit, murdering the clones in the process. But Merrick, too, comes down with the God disease; to justify his actions, he tells Albert Laurent (Djimon Hounsou), the bounty hunter who's tracking the clones Lincoln Six Echo and Jordan Two Delta, that he sees himself as a healer, not a murderer:

The Island of Dr. Moreau (1996). Nobel laureate scientist Dr. Moreau (Marlon Brando) has superb if mad self-confidence that he can genetically improve the human race, but in a classic example of overreaching, his efforts lead only to the desperately confused semihuman Beast People and utter chaos on his private island.

Source: New Line/The Kobal Collection.

> **Merrick:** I have discovered the Holy Grail of science, Mr. Laurent. I give life . . . The possibilities are endless here. In two years' time I will be able to cure children's leukemia. How many people on Earth can say that . . .
>
> **Laurent:** I guess just you and God. But that's the answer you want, isn't it?

The God virus also infects Michael Drucker of *The Sixth Day*. Though not a scientist himself, this wealthy businessman achieves his version of godhood by hiring scientist Griffin Weir to carry out cloning that can bring Drucker (and others) back from the dead.

In contrast to these stereotypes, nerdy to heroic to evil, are the fictional scientists who appear as rounded characters and show the motivations and traits of real scientists. Astronomer Ellie Arroway (Jodie Foster) in *Contact* is an example. The film opens as radio and television waves radiate outward from Earth to the rest of the universe at the speed of light, carrying programming from the 1930s until the present day. Then we focus on Ellie as a young girl, trying to communicate as far as possible with her ham radio set. Her father affectionately calls her Sparks, the nickname for radio operators. He is all she has, since her mother is dead, and he nurtures her scientific interests. Father and daughter do astronomy together, and she asks if there are people on other planets. Her father replies, "If it's just us, it seems an awful waste of space."

Along with ham radio and astronomy, that comment sets a theme that comes to dominate Ellie's life, as we see in a flash-forward to her as an adult at the enormous radio telescope at Arecibo, Puerto Rico. She has become an astronomer who detects radio waves reaching Earth from the rest of the universe, hoping to hear intelligible signals from an advanced alien race. Ellie listens to the cosmos with extraordinary intensity and determination. She persists even when the project loses its government funding and she must move to other telescopes in New Mexico, even in the face of advice that her research is out of the mainstream and will hurt her career.

Ellie's drive has deep roots. A flashback shows that her father died of a heart attack when she was eight. A Catholic priest tries to comfort her, saying that we're not meant to understand God's will, but she's unreceptive. She only regrets that she couldn't get her father's heart medicine to him in time. The early loss of her parents has created an inner loneliness that fuels Ellie's drive to reach out, but perhaps it has also limited her emotionally: when she meets handsome Porter Joss (Matthew McConaughey), a young missionary who questions the answers science can give, there's mutual attraction, and they have sex. But though he wants to further the relationship, she'd rather work on her astronomy.

As the story unfolds, Ellie actually discovers coded signals from beings many light-years distant. After an intervention from S. R. Hadden (John Hurt), a mysterious billionaire who has watched over Ellie for years, the signals are found to be plans for an advanced transport device that can carry a human to the aliens' home system. The government builds the machine; Ellie eagerly volunteers to be its passenger; but she is rejected because she refuses to say she believes in God, as foreshadowed in her childhood scene with the priest.

Contact (1997). Radio astronomer Ellie Arroway (Jodie Foster) listens tensely for aliens and later bravely volunteers to travel through the universe to their home. Though female scientists remain a minority in reality and on film, the way *Contact* and Foster portray Arroway is one of the more realistic screen images of a scientist, male or female. Her search for aliens, however, is hardly a standard career track. *A Golden Eagle; see chapter 9.*

Source: Warner Bros/Southside Amusement Co./The Kobal Collection.

Hadden, however, has secretly built a duplicate machine, and when the first one is destroyed by sabotage, Ellie enters the back-up model and is launched across the galaxy. Arriving at what seems a beautiful tropical island, she encounters the aliens, but in an unexpected way: they've tapped the contents of her mind and approach her in the guise of her father, so that her journey over light-years brings her back to her emotional roots.

As Ellie travels through wormholes and cosmic scenes in the alien machine, accompanied by spectacular visual effects, there is no doubt that *Contact* is science fiction, but within this framework, Ellie's life as a scientist is shown in believable terms. She looks smart and attractive but not unrealistically glamorous, and she dresses suitably as a scientist. Her childhood has persuasive notes as well. Many women who make it in science, business, and other male-dominated arenas credit supportive fathers who helped them develop confidence. It's also believable that Ellie might choose work over romance. Without reaching levels of madness, many scientists find their research as important to them as their relationships and defer serious connections until later in life. When they do marry, their spouses often find that their partner spends a lot of time in the lab. In that mold, Ellie seems happy with the personal choices she's made (although she does eventually reconnect with Porter). As a final character note, when Ellie refuses to deny her lack of religious belief even at the cost of her dream of finding aliens, she shows a degree of integrity that, it's easy to believe, enters into her science as well.

Moments of characterization appear for other fictional scientists, although less coherently and extensively than in *Contact*. In *The Day After Tomorrow*, Jack Hall is a victim of the hard-working scientist syndrome: he has a poor relationship with his teenage son, partly because of Hall's commitment to his research. In *Twister*, tornado hunter Jo Harding saw her father carried off by a tornado when she was a little girl, a memory that informs her commitment to understanding tornadoes and saving people's lives.

Other films show how driven scientists can be, as in *Good Will Hunting*. The film won Academy Awards for best screenplay written directly for the screen (by Matt Damon and Ben Affleck) and best supporting actor (Robin Williams). Its title character, Will (Matt Damon), has an innate brilliance that enables him to quickly learn subjects from art to biochemistry and that makes him especially creative in pure mathematics. A bona fide genius who was deeply scarred by an abusive childhood, Will is hardly a typical scientist; the film is largely about the efforts of psychologist Sean Maguire (Robin Williams) to help Will face his emotional problems.

Will has been sent to Sean by MIT mathematics professor Gerald Lambeau (Stellan Skarsgård), who understands how extraordinary Will's abilities are. Through Lambeau, we realize the pressures and drives operating on scientists at the upper reaches. He's a famous mathematician who has won the Fields Medal, the equivalent of the Nobel Prize for mathematics, awarded to researchers under forty.

You might think reaching this pinnacle would make Lambeau a contented man, but he's an intense, unsmiling presence who seems anything but happy. Compared to his less eminent old college classmate Sean, who teaches at a lowly community college, Lambeau's life is arid. Sean puts heart and soul into his psychological practice, loves baseball, and deeply loves and honors the memory of his dead wife; Lambeau interacts with his students through the brittle language of mathematics and shows no signs of a fulfilling personal life. His entire being is wrapped around further mathematical achievement, and if he can't himself become an even greater mathematician, he wants Will to do so, though Will wants to follow his own life.

Lambeau's characterization is a telling image of how love of the work, satisfying the ego, and a need to prove oneself can dominate a scientist's life and translate into the desire to pass on the passion and the obsession. Similar feelings come through in *Proof* (2005), from David Auburn's Pulitzer Prize–winning stage play of the same title. The story centers on Catherine (Gwyneth Paltrow), a young would-be mathematician whose father, Robert (Anthony Hopkins), was a famed mathematician of extraordinary brilliance. Robert made his reputation when he was young, but he later went insane, though he still has lucid moments. Catherine has given up her own mathematical aspirations and her social life to care for her father.

When Robert dies, his graduate student Hal (Jake Gyllenhaal) searches the 103 notebooks Robert has left and finds a major new mathematical proof that may have been derived by Robert—or by Catherine. As Hal and Catherine's sister, Claire (Hope Davis), interact with Catherine, and as Catherine talks to her father in flashbacks and inside her own mind, we see the intensity that drove Robert and his desire that his daughter share it. We also sense the thin border between madness and brilliance that Catherine has inherited from her father and see as well Hal's driving need to achieve his own mathematical success.

One real-life example of intense devotion comes from the British-born mathematician Andrew Wiles, who at the age of ten learned of the unsolved 300-year-old mathematics problem called Fermat's Last Theorem. After earning a doctorate and taking a position at Princeton, Wiles spent seven years—much of it isolated in the attic of his home—secretly working on a proof of the theorem, which he announced in 1993. But there was a mistake in the proof, so it was back to the drawing board for Wiles, who, with the help of a student (much as Lambeau in *Good Will Hunting* works closely with his graduate student), came up with a corrected proof in 1995. Wiles was slightly too old to win the Fields Medal, but he won another prestigious mathematics prize and received special mention at the Fields Medal ceremonies.

As the comparison between film mathematicians and Andrew Wiles's saga shows, a well-done fictional film can show something real about the psychology of scientists. These stories also reflect the real demographics of science; notwithstanding Ellie Arroway, there have been few female scientists in film. But more are appearing, according to recent studies carried out by Eva Flicker, a sociologist at the University of Vienna, and Jocelyn Steinke, of Western Michigan University. Flicker examined sixty films made between 1929 and 2003 that feature fictional scientists, and found that only eleven, that is, 18 percent, show female scientists. When Steinke examined films from a more recent period, however, 1991 to 2001, she found a higher percentage showing female scientists, twenty-three out of seventy-four films, or 31 percent, indicating that the representation of women is increasing on film as it is in real laboratories.

The quality of the representation matters, too, because even when a female scientist appears, she may play a secondary part. In *When Worlds Collide*, astronomer Cole Hendron has a main role in tracking the space objects that mean disaster for the Earth and in mobilizing the effort to save part of humanity. His daughter, Joyce, is described as an astrophysicist, but she does little more than assist her father and fall in love with the good-looking spaceship pilot. At least, however, she hasn't been made impossibly glamorous. In other films, gorgeous women are identified as scientists with little supporting evidence, leading to the suspicion that the title "scientist" is just meant to add a thin icing of pseudo-seriousness to a beautiful face and body. These honorary scientists include astrophysicist Charlie Blackwood in *Top Gun* and characters in the James Bond movies *Goldeneye* (1995) and *The World Is Not Enough* (1999).

But starting in the 1990s, stronger film roles for female scientists emerged besides Ellie Arroway, although she's the most fully formed character. In *Jurassic Park*, paleobotanist Ellie Sattler brings her own expertise to the story, shows bravery in dealing with the escaped dinosaurs, and at the same time dreams of having children with her scientist boyfriend. In *The Puppet Masters*, Mary Sefton is the lead biologist in examining the alien slugs and is confident enough of her sexuality to use it to sense which men have been taken over by them. In *Outbreak*, CDC physician Robby Keough leads a team to the town where the Motaba virus is raging, though she does succumb to the disease while her former husband, Sam Daniels, has all the adventures.

In *Twister*, Jo Harding is a dedicated and capable tornado hunter and shares the risks equally with her husband. In *Volcano*, Amy Barnes brings geological know-how and doesn't hesitate to investigate the dangerous situation under the streets of Los Angeles.

Still, stereotypes linger. Even in the 1990s female scientists could be downplayed, as can be illustrated by comparing *Chain Reaction* and *The Saint*. Each features a female scientist involved with fusion power, but they treat the two characters very differently. In *Chain Reaction*, although physicist Lily Sinclair has the Ph.D., technician Eddie Kasalivich figures out how to make the hydrogen energy process work. When Eddie and Lily go on the run, his physical abilities and ingenuity keep them ahead of their pursuers, though Lily helps. After they finally win free, it's Eddie who decides to give fusion power to the world. Lily emerges as having good looks, but not much else. She's mostly along for the ride as Eddie does the science, saves her, and makes the moral choice to give fusion power away; in short, she's the second banana, not so different from Joyce Hendron in *When Worlds Collide*.

The Saint, however, shows that a female scientist can be smart, sexy, and strong. As in *Chain Reaction*, the male lead's action scenes dominate the film, but unlike Lily Sinclair, Emma Russell is the scientific center. Moreover, she shows grit and moral courage. When she finds that the Saint has romanced her and had sex with her only to steal the secret of cold fusion, she seeks him out and confronts him. Although he wants to profit from cold fusion, she persuades him that the ethical thing is to give the secret to the world. Emma emerges as having it all: fulfilling work at which she's very good, intelligence, idealism, femininity and sexuality, beauty, and the Saint. (The fact that cold fusion is unworkable doesn't detract from Emma's portrait, though it certainly detracts from the film. Nor does every scene paint Emma in a serious way, such as one where she hides the formula for cold fusion in her bra, much to the Saint's excitement.)

Even the best of these fictional characters can't be painted with the same completeness as the real ones portrayed in biopics. *Kinsey*, *Gorillas in the Mist*, *A Beautiful Mind*, *The Insider*, and a handful of other dramas convey something true about scientists and their lives. They show some real heroes, as well as intensely committed scientists, illustrating that obsession isn't a purely fictional device.

Since a spate of scientist biofilms in the 1940s, however, the pickings have been slim for this genre, and even slimmer for female scientists. One of those 1940s films was *Madame Curie* (1943), about the double Nobel Prize–winning Polish-French scientist Marie Curie and her scientist husband, Pierre. Decades later, in 2003, the BBC produced the television documentary *Rosalind Franklin: DNA's Dark Lady,* about the British X-ray crystallographer whose work helped decipher the structure of DNA. Another relatively recent biopic about a female scientist, made to full Hollywood scale, is *Gorillas in the Mist* (1988), about primatologist Dian Fossey (Sigourney Weaver).

As the film points out, Fossey wasn't trained in that area. First studying to be a veterinarian, she eventually became an occupational therapist who related unusually well to disabled children—an experience that prepared her to work with gorillas. As the film opens, anthropologist Louis Leakey (Iain Cuthbertson), famed for his work on human origins, is hiring Fossey to study mountain gorillas in Africa. Leakey believes that studying these primates will shed light on the primate fossils he's been finding in Africa.

In 1966, Fossey leaves behind her life and fiancée in the United States, and sets up camp in the Democratic Republic of the Congo (called Zaire between 1971 and 1997). Persisting even through civil war, she becomes fully accepted by the gorillas and is the first human to touch one. As she becomes involved, she becomes upset that the small number of gorillas is further reduced by poachers who abduct them for zoos or kill them to turn them into tourist items. She becomes fiercely committed to preserving the animals by any means: she pretends to be a witch to frighten superstitious natives, publicly confronts a collector for a zoo, and shoots at tourists encroaching on "her" mountain. This zeal and her growing isolation bring her to the edge of madness, but it all comes to an abrupt end when, as actually happened in 1985, Fossey is discovered murdered in her cabin. The killer is never found, though it's probably a poacher who resented her efforts.

Dian Fossey's science went beyond an objective interest in her subject: she loved the mountain gorillas, which in a way led to her death. It's understandable that a scientist can become emotionally attached to animals under study, but the film doesn't show what in Fossey drove her toward her deep connection with her gorillas and her isolation from people. In any case, compared to scientists who study living creatures or humans, physical scientists are less likely to become emotionally involved with their subject, though they can bring equal passion and curiosity to their research.

That passionate curiosity is shown in *Infinity* (1996), about physicist Richard Feynman (Matthew Broderick). The film didn't make a major splash when it was released, but it provides a portrait of an important contemporary scientist. Feynman, who died in 1988, was a brilliantly intuitive physicist; he won a Nobel Prize for his work in the quantum theory of light and was instrumental in investigating NASA's *Challenger* shuttle disaster in 1986. The film begins as six-year-old Richie Feynman (Jeffrey Force) asks his father Mel (Peter Riegert) endless questions about everything around them. His father answers carefully, even presenting the idea of inertia to explain how a ball moves in the red wagon Richie is pulling, and encouraging his son much as Ellie Arroway's father does in *Contact*. In voiceover, the adult Feynman pays tribute to his father, who was not a scientist: "Not having experience with many fathers, I didn't realize how remarkable he was. How did he learn the deep principles of science and the love of it, what's behind it and why it's worth doing? I never really asked him, because I just assumed those were things that fathers knew."

The film moves on to Feynman's teenage years, when he becomes smitten with Arline Greenbaum (Patricia Arquette), who sings and paints. They become a couple as Feynman graduates from MIT and enters Princeton for a physics Ph.D. Meanwhile, World War II breaks out, and Feynman is invited to join the Manhattan Project to build an atomic bomb. Things take a dark turn, however, when Arline develops tuberculosis, which in the 1940s is extremely serious, though not necessarily fatal.

As Feynman works at Los Alamos and visits Arline—now his wife—at a hospital in Albuquerque, the film shows their warm relationship along with further signs of Feynman's curiosity about the world, for instance, his tests of the capacities of the human sense of smell. The film also shows his occasional arrogance and disdain, as in a scene where he flaunts his success in beating what he calls the "exaggerated and at the same time . . . absolutely ridiculous" security system at Los Alamos. Interspersed with these personal moments are reminders of what's going on at Los Alamos, in flash-forwards where Feynman and others don dark goggles to prepare for what we know will be the first atomic bomb test.

Two years go by, and the Allies declare victory in Europe. Soon after, Feynman learns that Arline is in a bad way. He frantically drives to Albuquerque and arrives in time to exchange a few words at her bedside before she dies, though his reactions seem muted. Not long after, we see Feynman at the Alamogordo test site as the A-bomb is detonated. He gazes awestruck at its power, followed by an immediate flashback to six-year-old Richie and his father talking about the ball in Richie's wagon. The final scene shows Feynman's emotional response to Arline's death. However cool his reaction right after she died, when he unexpectedly sees a red dress in a shop window that reminds him of a dress she once wanted, he finally cries for Arline.

Though *Infinity* focuses on Feynman's relation with Arline, the film refers to Los Alamos and to the moral issues the atomic bomb raised. *The Insider* (1999), nominated for seven Academy Awards, also involves a real scientist and displays a different question about a scientist's ethical responsibilities. The scientist is Ph.D. biochemist Jeffrey Wigand (Russell Crowe), vice president of research and development for the Brown & Williamson tobacco company. The position pays well and provides the good life for Wigand and his family. But Wigand is troubled when he finds that the company processes its cigarettes to quickly deliver the maximum nicotine hit—although the CEOs of Brown & Williamson and the other six biggest tobacco companies had publicly sworn before Congress that cigarettes are not addictive.

In 1995, Wigand is approached by Lowell Bergman (Al Pacino), a producer for the CBS TV show *60 Minutes*, and agrees to tell what he knows on air. The film shows how *60 Minutes* handles the story and how CBS management decides not to air Wigand's interview for fear of lawsuits. Bergman emerges as a hero of ethical journalism, and Wigand as a hero of ethical corporate science. He knows that the drug and health industries have higher standards than Big Tobacco. In a pivotal scene, Wigand tells Bergman how the CEO of Johnson & Johnson once acted quickly and responsibly to remove a tainted product from the shelves because the CEO was a "man of science" as well as a businessman. Wigand, too, sees himself as a "man of science" and feels guilty about acting as a hired gun for a company whose product hurts people. He goes ahead with his revelations despite serious personal costs: losing his job and taking a less lucrative position as a high school teacher, facing possible legal action, having his wife leave him, and receiving death threats.

There are other real situations where devotion to scientific truth leads to a form of heroism, as in *Kinsey* (2004), about Alfred Kinsey, the first scientist to methodically study human sexuality, beginning in the 1930s. The film is unusual in presenting a social scientist, and a scientist whose work was controversial and remains so in some quarters; not everything the film presents as biographical fact is universally acknowledged as true or correct. Besides the sensational nature of Kinsey's research, there are subjective elements that wouldn't enter in the same way for a physicist. One is the relation between Kinsey's upbringing and personality and his crusade for valid research into human sexuality that would dispel sexual ignorance.

The film begins as we watch the adult Kinsey (Liam Neeson) train a colleague in the interview methods Kinsey developed to probe the sexual history of volunteers, his main scientific tool. To illustrate the influence of Kinsey's background, the film intercuts scenes from his past. Some come from his childhood, showing how the young Kinsey (Kinsey at fourteen is played by Matthew Fahey, and at nineteen by Benjamin Walker) was raised under the thumb of a rigidly religious, domineering father (John Lithgow) who tried to squelch every sensual impulse. Other background scenes come from Kinsey's marriage to his former student Clara McMillen (Laura Linney), which initially had sexual difficulties.

Tracing Kinsey's development, the film shows him breaking away from his early influences to earn a Harvard Ph.D. in zoology and then join the faculty at Indiana University. At first he works on a monumental study of the gall wasp, but after teaching a course on marriage, he begins examining sexuality. He compiles the results of thousands of interviews about men's sexual histories in *Sexual Behavior in the Human Male*, published in 1948. The book is greeted with headlines such as "Kinsey Drops His Atom Bomb" and catapults Kinsey to national fame. His studies explore long-hidden areas and upset conventional wisdom about sex; it shows, for instance, that extramarital sex and supposedly perverse sexual activities such as masturbation and homosexuality are far more widespread than anyone had known.

Despite Kinsey's provocative subject, he's shown as professorial in style and conservative in appearance, with short hair and an ever-present coat and bow tie. His own sexual life seems anything but conservative, however; according to the film, it includes bisexuality, masochism, and extramarital sex for himself and his wife, all graphically portrayed. But whatever his personal life, Kinsey makes his science as solid as possible; as he tells a colleague in the movie, "without measurements, there can be no science." No hard-core physicist could have put it better.

This reliance on the factual, though, leads to a blind spot for Kinsey that can trap any scientist devoted to rationality, namely, neglecting the power of emotion. Emotion can override reasonable arguments, say, about the theory of evolution or about the safety of nuclear power, and even more so when the subject is sex, but the film shows Kinsey failing to understand why his work isn't always accepted. Strong protests greet his second book *Sexual Behavior in the Human Female* (1953), called "an insult to women" in which Kinsey "insecticizes American womanhood." Still, Kinsey continues his pursuit of facts. If his efforts offended some, they made him a hero to others, like the lesbian in the movie who thanks him for his work, which gave her the courage to follow her impulses and find happiness with another woman.

Scientific brilliance, intensity, and dedication can lead to heroic acts, but if these qualities are allied to madness, they have a dark side as well. In the romantic tradition, great artists and writers supposedly walk a narrow line between sanity and madness. The same may be true for creative scientists. But although fictional scientists from Dr. Frankenstein to Catherine in *Proof* are portrayed as living on or near that boundary, there are few real examples suggesting that madness has any connection to scientific ability.

In *A Beautiful Mind*, however, which won an Oscar for best picture of 2001, we meet John Nash (Russell Crowe), a real mathematician who became clinically insane. Nash shared the 1994 Nobel Prize in Economic Sciences with two other laureates, an award that came after he had struggled with schizophrenia for decades. The film is based on Sylvia Nasar's (unauthorized) 1998 biography of Nash, of the same title, but takes many liberties with the book. Nevertheless, it conveys much about the general shape of Nash's mathematical brilliance and of his life. The story begins in 1947, when Nash, with other mathematics graduate students entering Princeton, hears Professor Helinger (Judd Hirsch) explain that mathematicians helped win World War II and that they will continue acting as "the vanguard of democracy" in the Cold War.

We see more of Nash as he displays a breathtaking lack of social skills and spectacular arrogance, telling a fellow student that his writings contain "not a single seminal or innovative idea." Nash fully understands that he doesn't deal well with people; he tells his roommate Charles Herman (Paul Bettany),

"My first grade teacher once said I got two helpings of brain and half a helping of heart," adding, "I don't like people much and they don't much like me." But his reputation as an odd duck, and his lack of success with women, is less important to him than finding an original mathematical idea.

After much struggle, he develops an idea so good that it's a major breakthrough in the theory of games, the mathematics of how people interact in cooperative or noncooperative ways. As Nash explains, the optimal collective strategy among competitors is one where each competitor makes choices based on the expected rational behavior of all the others. Nash illustrates his idea in a scene where he and his fellow students approach a group of young women in a bar. Nash says that if all the men focus on the best-looking woman, their efforts will cancel out with no happy result for any of them; but if each man talks to one of the other women, each will end up in bed with a woman, giving the best overall outcome.

This insight about group behavior earns Nash a coveted job in a leading government defense laboratory at MIT. He quickly shows his brilliance at code breaking, an essential weapon in the Cold War. Soon he's approached by government agent William Parcher (Ed Harris), who recruits him for additional secret undertakings involving a Soviet A-bomb in the United States and coded messages in newspapers. Meanwhile, Nash also meets Alicia Larde (Jennifer Connelly), a student in one of his classes, who loves and accepts him, eccentricities and all.

They marry, but then Nash begins behaving strangely, believing that he's pursued by Soviet agents. Finally psychiatrist Dr. Rosen (Christopher Plummer) diagnoses Nash as suffering from schizophrenia that induces paranoid behavior, which had affected him even in graduate school. His supposed roommate and friend Charles exists only in Nash's troubled mind, as does Charles's young niece Marcee (Vivien Cardone) and the mysterious government agent Parcher.

After devastating shock treatments and drugs, Nash achieves a fragile balance. Alicia copes with his condition and a new baby, while Nash tries to continue his mathematics but often relapses into schizophrenia. In time, though, his mental state improves, partly as he realizes the logical inconsistencies in his delusions: for instance, that the illusory Marcee never gets older. He begins teaching again, interacting with and inspiring Princeton students. The culmination of his struggles and the validation of his ability come when he receives a

A Beautiful Mind (2001: four Academy Awards, including best picture). Mathematician John Nash (Russell Crowe) in a moment of madness and mathematics at Princeton. Like another scene illustrating the mathematical idea that won the real Nash a Nobel Prize despite years of schizophrenia, the symbols on the window actually make sense—at least, to another mathematician. Biographical films about scientists generally show their science. *A Golden Eagle; see chapter 9.*

Source: Dreamworks/Universal/The Kobal Collection/Eli Reed.

Nobel Prize that recognizes his work in game theory. The idea he illustrated with students approaching a group of women is now called the Nash Equilibrium and has impact in areas from trade negotiations to evolutionary biology.

Any biopic requires picking and choosing among the elements in a person's life to fit into a brief two hours. *A Beautiful Mind* goes further: it alters Nash's life and the details of his disease to make the story more dramatic and emotionally compelling. For example: Nash never worked on code breaking. He did marry a physics student at MIT, but contrary to the story, she divorced him after he was diagnosed with schizophrenia, although they later remarried. And though Nash experienced hallucinations, they were auditory—voices in his head—rather than visual (schizophrenics can suffer from both, but auditory hallucinations are more common).

In any case, with his descent into madness followed by a Nobel Prize, Nash is hardly a typical scientist; yet precisely because of his harrowing life, his story compellingly illustrates the power of that enduring scientist characteristic, "perfection of the work." And although his fine mathematical mind was no proof against schizophrenia, we're reminded that rigorous thinking and reliance on fact is the opposite of chaos and madness—or is it? Like the fictional situation in *Proof*, Nash's story carries us along that supposedly thin line separating madness from genius, though it doesn't answer the continuing question, are the two linked? Perhaps nobody can provide an answer; but in his own Nobel Prize autobiographical statement, Nash gives some insight into the connection, saying:

> At the present time I seem to be thinking rationally again in the style that is characteristic of scientists. However this is not entirely a matter of joy. . . . One aspect of this is that rationality of thought imposes a limit on a person's concept of his relation to the cosmos. . . . [One] could think of Zarathustra as simply a madman who led millions of naive followers to adopt a cult of ritual fire worship. But without his "madness" Zarathustra would necessarily have been only another of the . . . billions of human individuals who have lived and then been forgotten.

Nash's unusual story still fits within the spectrum of movie depictions of scientists. Some films display cardboard cutouts, some show glimmers of understanding of the scientific temperament, and some show people who happen to be scientists realized as fully as a two-hour movie can achieve.

We've also looked at film depictions of science. Some, as in *The Core*, simply seem uncaring about getting the science right; some, as in *The Day After Tomorrow*, start with valid science but distort it, more or less, to enhance drama and meet cinematic constraints; and some, like *Fat Man and Little Boy*, make an effort to follow the real science accurately and present it in some detail. So having looked at film science and film scientists, now we can answer two questions: what makes a good science-based film, and what makes a poor one?

Chapter 9

Solid Science and Quantum Loopiness

Golden Eagles and Golden Turkeys

> **Replicant Roy Batty:** I've seen things you people wouldn't believe. Attack ships on fire off the shoulder of Orion. I watched C-beams glitter in the dark near Tanhauser Gate. All those moments will be lost in time, like tears in rain. . . . Time to die.
>
> —*Blade Runner* (1982)

> **Dr. Josh Keyes:** I'm married to my work.
> **Serge Leveque:** So am I. Which makes my wife my mistress. That's why I'm still in love with her.
>
> —*The Core* (2003)

When the Oscars are handed out every year by the Academy of Motion Picture Arts and Sciences as millions of people watch, there are no special awards for "best portrayal of a scientist" or "best explanation of quantum theory." Still, a few films that have some relation to science have enjoyed serious recognition. In 1998, *Good Will Hunting* was nominated for several awards, including best picture, and took an Oscar for best original screenplay, with another Oscar for Robin Williams as best supporting actor. *A Beautiful Mind* won four major Academy Awards in 2002: best picture, best director (Ron Howard), best screenplay from previously published material (Akiva Goldsman), and best supporting actress (Jennifer Connelly).

Other films with a scientific or futuristic bent, mostly science fiction, have won Oscars, too, though not in the top categories. *Destination Moon* began a trend when it won in 1951 for its special effects, and some films have had chances to reach higher. After *Dr. Strangelove* was nominated for best picture in 1965, *2001: A Space Odyssey*, *A Clockwork Orange*, *Star Wars*, *Close Encounters of the Third Kind*, and *E.T.* were nominated for best picture or director awards, but all took home only lesser Oscars for musical score, cinematography, or—inevitably, following in the footsteps of *Destination Moon*—special or visual effects.

Superior science fiction films, however, have been recognized in other ways. In 1998, the American Film Institute named eight of them to its list of the one hundred all-time best American films (or eight and a half, if you add *Bringing up Baby*, which is hardly science fiction but does have Cary Grant playing a paleontologist). There's even a science fiction Oscar of sorts; it's the Saturn award, presented since 1972 for the best science fiction film of the year by the Academy of Science Fiction, Fantasy and Horror Films (www.saturnawards.org). Any list of these bests includes modern classics such as *Close Encounters of the Third Kind*, *Star Wars*, and *2001*.

To balance the earnest search for good films, there are snarkier approaches that delight in ripping apart bad films, a depressingly large number of which fall in the science fiction category. The 1980 book *The Golden Turkey Awards* by Michael and Harry Medved invented those imaginary prizes for some less-than-wonderful films. Another source is the Bad Cinema Society, which presents its list of "ultimate bad movie awards," as determined by popular vote, on the Internet (www.listsofbests.com/list/69). In a parody of the hundred best films from the American Film Institute, this site lists the hundred worst films of the twentieth century, including a dozen science fiction efforts. As if this weren't distinction enough, *The Golden Turkey Awards* considers *Plan 9 from Outer Space* (1959) not just the all-time worst science fiction film, but the all time worst *film*, period. Another more recent science fiction flick, *Battlefield Earth: A Saga of the Year 3000* (2000), wins the same title from the Bad Cinema Society.

Awards and lists that rate films according to the usual cinematic criteria—screenplay, directing, acting, production values, music, and so on—don't necessarily consider all the features that might enter into an exceptional science film or are lacking in a poor one. Along with cinematic power, these include reasonable scientific veracity; a meaningful portrayal of a scientist or scientists; an effective look at how science affects people and society; and a persuasive and thought-provoking extension of today's science into tomorrow's world.

Some analyses do focus on one of these traits, scientific accuracy. The Web site Insultingly Stupid Movie Physics (www.intuitor.com/moviephysics) rates the physics content of movies, from *The Terminator* to the latest installment of the Star Wars saga on a five-level scale that runs from GP or "good physics," through PGP or "pretty good physics, with just enough flaws to be fun," to XP or "physics from an unknown universe." The categories are amusing, but the dissection of the use and misuse of physics concepts is well done.

However, if we were to insist on total scientific accuracy, many films would fail because some plots simply can't work unless science is extended in unlikely or impossible directions. The classic example is faster-than-light (FTL) travel. As we've known since Einstein's Special Theory of Relativity came along in 1905, no object in the universe, from a spacecraft to a star, can travel at or beyond the speed of light, 186,000 miles (300,000 kilometers) per second. This would put a terrible crimp in stories like *Star Wars* or *Star Trek*, where humans reach the stars, or that involve galaxy-wide civilizations, because distances in space are so enormous. Even at the speed of light, let alone the much lower speeds that can actually be attained, the travel times for spaceships going from Star A to Star B in the galaxy would range from years to thousands of centuries.

In science fiction, this inconvenient fact is finessed by imagining a spaceship engine that somehow alters space and time (frequently called a hyperdrive, or in the *Star Trek* series, a warp drive) to propel the ship at multiples of the speed of light. In *Star Trek*, for instance, a speed of "warp 5" means the *Enterprise* is moving at 214 times the speed of light. At this speed, it could reach Alpha Centauri, 4.4 light-years away and our closest star other than the Sun, in just over a week. Similarly, in *Star Wars*, when Han Solo's fast spacecraft the *Millennium Falcon* flashes into high gear, the stars look distorted through its viewport. This is an imagined representation of the universe as seen when traveling beyond light speed.

FTL travel is a convention that allows action to happen in reasonable time spans across vast spaces. Other conventions also venture beyond known science into arenas where scientists have barely a clue. These include FTL travel across the universe via shortcuts through wormholes—that is, distortions in spacetime predicted by Einstein's relativity—which is Ellie's mode of transport in *Contact*; travel through time as in *Back to the Future*, also sometimes associated with wormholes; matter transmission through space, which happens whenever Scotty beams people from the *Enterprise* to a planet; antigravity, implied by the ease with which an unbelievably huge spaceship hovers over the White House in *Independence Day*; and force fields, as used by the Martians in the original *The War of the Worlds* to deflect an atomic bomb and by the aliens in *Independence Day* to protect their spaceships.

None of these possibilities is valid science, at least not yet, but we the viewers suspend our disbelief and accept the convention for the sake of the story. The result has been some of the best and most popular science fiction films, although these are rarely the ones with the greatest scientific content or the deepest meaning for science and society.

In any case, scientific accuracy is only part of an evaluation. We might also ask, does the film show something about the scientific mind and style? Does it explore relations between science and society, government, the military, or the corporate world? Does the film treat ethical issues arising in science itself or from the uses of science? Does it draw attention to important problems that science creates or that science can solve? If the film speculates about science, now or in the future, does it invite the viewer along and stimulate his thinking?

Even some significant science-based films don't manage to unite great filmmaking with thoughtful considerations of science. Some aren't even very good movies, like the classic *Destination Moon*. Robert Heinlein (with two other writers) adapted the film from his novel *Rocketship Galileo* and acted as technical adviser. As in all his writings, he emphasized scientific correctness. The spaceship, and conditions in space and on the Moon, are depicted according to the best knowledge of the time. But the plot is charitably described as less than gripping, the dialogue is flat, and the characterization paper-thin. Still, the film is important because it dares include scientific exposition, which it lightened by using Woody Woodpecker to show principles of space flight.

Other films show less about science but more about scientists, or science in society. *Infinity* says little about the ideas of physics but gives insight into the mind and personality of Nobel laureate physicist Richard Feynman and alludes to the atomic bomb and its impact. *The Sixth Day* makes sharp comments about science and religion as it explores responses to human cloning. It also cleverly projects today's technology forward, with aircraft that fly by remote control and smart refrigerators that know when food is running low. But since the film is devoted to Arnold Schwarzenegger's action scenes, there's little time to flesh out the scientist character, Dr. Griffin Weir. A bigger problem is that the core idea driving the plot, the instant duplication of an adult by cloning, is flat wrong. In pushing the idea of cloning far, far beyond possible or even remotely believable limits, *The Sixth Day* loses its basic nugget of science.

These examples don't fully meet the criteria of good science, well-portrayed scientists, social impact of the science, or a convincing future—all set within a film that people actually want to see. Nevertheless, the categories give a useful framework against which to select some of the best and the worst science-based films. Here are my own Golden Eagles and Golden Turkeys of Science, along with one Special Award, followed by my reasons for choosing them:

GOLDEN EAGLES

Metropolis (1927)
The Thing from Another World (1951)
The Day the Earth Stood Still (1951)
On the Beach (1959)
Blade Runner (1982)
Jurassic Park (1993)
Gattaca (1997)
Contact (1997)
A Beautiful Mind (2001)

SPECIAL AWARD

The Day After Tomorrow (2004)

GOLDEN TURKEYS

The Core (2003)
What the #$!Do We (K)now!?* (2004) (alternate title: *What the Bleep Do We Know!?*)

Golden Eagles

Metropolis (1927). Apart from other influential qualities of Fritz Lang's seminal film, *Metropolis* belongs on the best list for its audacious projections of technology and for its early depiction of connections between technology and public affairs. The city's towering buildings, underlaid by huge underground machines that power and control everything, are a technological vision. This reaches a peak in Rotwang's robotic woman, far beyond the scientific capability of 1927 and even today. Technological advance is also expressed in details like the ultramodern spiral-shaped neon lamps illuminating Rotwang's laboratory and Joh Frederson's two-way video communicator. Both neon lighting, called "liquid fire," and television, were only in their infancy in 1927.

The film goes beyond technical achievements to hint at how technology affects society. The easy high-tech life of the upper classes requires hordes of machine tenders, but Rotwang imagines a world of robot laborers, implicitly eliminating the human workers—a foretaste of today's concerns about replacing people with computers and robots, as is already happening in automobile manufacturing. When Joh Frederson secretly directs Rotwang's robot to blunt the worker's revolution, however, he's covertly using technology to support a political end, the preservation of the privileged classes. The relations between science and government, whether a secret collusion, a willing and open marriage, or downright hostility, appear in film after film thereafter, as they do in real life.

The conventions of the silent film look quaint today, and the only answer *Metropolis* gives to resolve the tension between the haves and the have-nots is to plead for good will in its tag line: "The heart must mediate between the head and the hands." But in its restored form with the original musical score, the film's spectacular cityscapes, crowd scenes, and stunning robot still give *Metropolis* considerable power.

The Thing from Another World (1951), *The Day the Earth Stood Still* (1951), *and On the Beach* (1959). Unlike *Metropolis*, each of these films from the science-fiction-rich 1950s is set in its own present time or shortly after, an era filled with the Cold War threat of global nuclear exchange, the biggest scientifically based issue of the time. Nuclear realities are implicit in *The Thing* and explicit in the other two films.

Although *The Thing* is mostly about an encounter with an alien, it comments on the nuclear age and the scientists responsible for it through a bystander's sardonic response after Dr. Carrington extols nuclear fission as a great achievement. Also, Carrington is ready to sacrifice lives to obtain the Thing's knowledge, while the military men are desperate to kill the creature, never mind the knowledge. This difference between a scientist remote from human concerns and men of action who know their duty further underlines the evils of science. However, unsympathetic though Carrington is, he does explain how the plantlike Thing can be mobile and intelligent, giving the film points for scientific exposition. But this doesn't slow the action, which unfolds at a rapid pace. The story is tautly told and has moments of horror that keep viewers enthralled.

The Day the Earth Stood Still has equations on a blackboard but no scientific discussion, though it displays advanced alien science that can disrupt electricity and raise the dead. Nor does the film feature nuclear explosions, but the alien Klaatu has come to Earth because of our entry into the nuclear age. As he confronts the fear and mistrust directed at him and among the nations of Earth, his message that we must control our nuclear activities becomes ever more meaningful. Of course, it's really a message from ourselves to ourselves to act wisely before it's too late.

Klaatu is helped by theoretical physicist Jacob Barnhardt, who resembles Einstein in his benign appearance and worldwide reputation. Immediately accepting the weight of Klaatu's mission, Barnhardt gathers scientists to hear the alien's message. Barnhardt is the opposite of Carrington in his concern for humanity and so partly redeems what the nuclear physicists have done in creating the atomic bomb.

This black and white film has only modest special effects, though Klaatu's space saucer and his robot Gort are effective (also an electronic theremin provides an unusually eerie, unworldly musical score that adds a great deal to the film). Instead, the film makes its points through a human-scale story in which Klaatu takes on the (Christlike?) role of Mr. Carpenter and interacts with ordinary people as well as scientists and government officials. Klaatu's reactions to human ways deliver the movie's message without excessive speechifying, and events move along at a satisfying clip.

In *On the Beach*, it's too late for wisdom: after a nuclear war, humanity's last remnants await their fate as a deadly cloud of fallout approaches. This movie also displays little science but does have physicist Julian Osborne. Neither benign like Prof. Barnhardt nor remote like Dr. Carrington, he's angry and rueful as he talks about how scientists contributed to the nuclear disaster, from Einstein's ideas to his own work on weaponry. In partial compensation, he goes along as the nuclear submarine *Sawfish* seeks a radiation-free zone, just as the submarine itself, a weapon driven by nuclear fission, ironically becomes humanity's only possible salvation.

The impact of impending death is shown on the human scale in the love affair between submarine Captain Dwight Towers and Moira Davidson, in the responses of a young Aussie naval lieutenant with his wife and baby daughter, and in the behavior of the submarine's crew. The total effect is of a world defeated, without the will to save even a few.

The movie's black and white format helps set a mood of resigned despair, and the star power of Gregory Peck and Ava Gardner adds weight to their love story. The viewer might get tired of hearing "Waltzing Matilda" in the background, but the melody is used to good emotional effect. The later TV version of the story brings Towers and Moira together for their last hours and humanity's. But the ending of the original version is far more chilling as Moira, alone in Australia, and Towers, aboard his submarine, both go to their fates. The bleak sense of their apartness, and the final shots of empty, desolate streets in Melbourne, still leave an audience stunned and mute as the film ends.

Blade Runner (1982). *Blade Runner* also shows little scientific detail, but it presents a compelling vision of artificial beings, set against a background that updates *Metropolis*. That city, with its lofty aeries and underground warrens, becomes a future Los Angeles dominated by the tower of the Tyrell Corporation, with lower levels populated by the poor and the criminal. Rotwang's robot becomes a whole technology and economy of replicants, human-seeming artificial beings that, like humans, are aware of their mortality. Knowledge of death spiritually separates humans from animals. That same knowledge brings replicants like Roy Batty closer to humanity, and perhaps closer to possessing a soul, than other artificial creatures in the movies.

If there's a great speech in science fiction cinema, it's Batty's final words to his nemesis, blade runner Rick Deckard, as Batty's life span runs out:

> I've seen things you people wouldn't believe. Attack ships on fire off the shoulder of Orion. I watched C-beams glitter in the dark near Tanhauser Gate. All those moments will be lost in time, like tears in rain . . . Time to die.

The speech underlines the replicant's humanlike characteristics mixed with its artificial capabilities. There's further interplay between the human and the artificial in Deckard's love for the replicant Rachael and in the possibility that Deckard himself is a replicant. *Blade Runner* distinctively combines a compelling future environment and technology, unique artificial beings, and a memorable mood to explore the biggest question of all: what does it mean to be human?

Jurassic Park (1993). After the physics-dominated films of the 1950s onward, *Jurassic Park* represented a breakthrough into biological science, matching its growing scientific weight and the increasing public interest in the area. Earlier films such as *Panic in the Streets* and *The Boys from Brazil* explored exotic diseases and genetic issues, but *Jurassic Park* provided a strong combination of a novel idea, scientific exposition, and scientists in action, along with revolutionary special effects.

Although the details of creating a living dinosaur from old DNA held in mosquito blood aren't realistic, the film does show the basic science of cloning in understandable terms. And although scientists Alan Grant and Ellie Sattler aren't presented in depth, we see something of the commitment that scientists can bring to their work, even at an emotional level; for instance, when Alan nearly faints at seeing the living object of his years of effort, a re-created *Tyrannosaurus rex*. Ellie, in her sympathetic desire to help a suffering dinosaur by tracking down the source of its illness, doesn't hesitate to plunge elbow deep into a huge pile of dinosaur droppings.

As an enormous bonus, not only do the dinosaurs nearly overcome Alan, they almost overcome the viewer. Their natural-looking portrayal is a triumph of special effects using new developments in computer-generated imagery (CGI). Without them, the movie could not have such a major impact—and the impact was great, showing viewers the immense possibilities of modern genetics.

Gattaca (1997). *Gattaca* also examines the impact of genetic engineering, but on people, not dinosaurs. Set in the "not too distant future," it tells a powerful story about one man's vision within a society based on genetic manipulation. The scientific content isn't detailed, and it's not even completely correct; the urine testing that plays a big role doesn't actually reveal DNA signatures. Nor does the film feature a scientist, except for a brief appearance by a geneticist.

But *Gattaca* explores the human outcomes of genetic control through its characters. Vincent is a sympathetic young man who draws us in with his dream of surpassing his supposed genetic limits. Jerome, a superb genetic specimen, is bitter about his career-ending injury yet also becomes caught up by Vincent's efforts. Vincent's love interest, Irene Cassini; his genetically superior brother; Dr. Lamar, whose son's genetic heritage didn't turn out as predicted; a genetically manipulated twelve-fingered pianist—these and other characters show what genetic manipulation would mean to individuals, families, and society.

The look of the movie suggests what drives this future world. Unlike *Blade Runner*, *Gattaca* lacks dramatic urban vistas, but the spotlessly sleek premises of the Gattaca Corporation show commitment to high technology, as is also reflected in the instantaneous DNA testing, quiet electric cars, and solar power in the film. Vincent is sent over 700 million miles to Saturn's moon Titan in an ordinary business suit, not a spacesuit; Gattaca routinely launches up to a dozen space shots a day. In this world of superb technology, Vincent's humanity shines through in Michael Nyman's lyrical score, which brings expressive emotion to the story.

Many small touches add texture. Irene is the namesake of Jean-Dominique Cassini, the seventeenth-century Italian astronomer who first saw some of Saturn's moons, where Vincent is bound. The twelve-fingered pianist plays Schubert's Impromptu for Piano in G-flat, Op. 90, No. 3, especially embellished for the film. The spiral staircase in Jerome's home reminds us of DNA's double helix, along with the name Gattaca. And there's the dual use of hair: On the one hand, it's no more than a scientific sample that can be rigorously tested to reveal a person's DNA. But hair is also a deeply human remembrance: just before Jerome decides to end his life, he gives Vincent a lock of his hair to take to Titan.

Gattaca inspires thought long after it's over. Without overt preaching, it shows subtly but unmistakably what it would have meant for the Nazis to successfully breed a super race, or how it feels to belong to a stereotyped minority. The movie goes beyond genetics and science to raise bigger questions about societal acceptance and rejection and what it means to be human.

Contact (1997). *Contact* is partly a science fiction tale about aliens, but it shines as a story that's more realistic than most about how scientists think and work. It also tries, though not wholly successfully, to raise deeper issues about science and faith.

Radio astronomer Ellie Arroway is a first-rate scientist whose search for alien messages has personal roots. She may seem overly obsessed, but no more so than many real scientists, though few would pursue her goal of finding aliens. She works hard and jumps through the right hoops, from earning a Ph.D. to convincing the National Science Foundation to fund her research. Then she faces what many actual scientists do: fashions in funding change, and she loses support for her project. In essence, the same thing happened in real life: SETI, the project that seeks radio signals from extraterrestrial intelligences, once received federal dollars from NASA. In 1994, however, Congress decided the research was too far out and pulled the support; now SETI is funded by a variety of sources.

Contact says true things about scientists as it shows how Ellie's childhood flowers into a scientific career and outlines her experiences in the research community. The second half of the film, in which Ellie voyages to a distant star, is less realistic, and although it tries to integrate Ellie's scientific mindset with spiritual and religious views, it doesn't do so convincingly. Still, her depiction, as written, and as acted by Jodie Foster, is among the better portrayals of a scientist in film—and a female scientist at that, an underrepresented category.

A Beautiful Mind (2001). *A Beautiful Mind* might seem an odd choice because of its unusual subject, a mathematician who goes insane. Its multiple Oscars attest to its cinematic qualities, but how about its scientific qualities? They come through in the portrayal of John Nash that shows traits representative of him as a scientist: his focus on finding the one crucial problem that will stretch his abilities, his determination to keep working as a creative mathematician despite his illness. And although Nash's connection with government code breaking is fictional, this kind of association is plausible. In that Cold War era, U.S. scientists contributed to military and security operations, often through such organizations as the RAND Corporation. Finally, *A Beautiful Mind* goes the extra mile by showing the abstract idea that won Nash the Nobel Prize in a way that fits into the story, through the scene where graduate students approach a group of women.

It's difficult for a film to convey the essence of a scientist's contribution, especially for ideas of theoretical physics or mathematics. *Proof*, another story about mathematicians, only tells us that the mathematical theorem in question is highly significant; we're given no details to back up this assertion. In *Infinity*, although we see Feynman's curiosity about the natural world, there is hardly an inkling of his original thinking in physics that led to a Nobel Prize. But in *A Beautiful Mind*, we actually see something of the idea that made Nash famous.

Special Award

The Day After Tomorrow (2004). *The Day After Tomorrow* didn't receive much critical acclaim when it was released. It wasn't on anyone's Oscar list, either, though it received lesser awards from other organizations for its visual effects and sound and for two of its young actors, Jake Gyllenhaal and Emmy Rossum. But the plot and dialogue were thought to be trite and predictable.

Still, the film has redeeming features. Though presented in clichéd style, the story illustrates scientific commitment through the conflict between Jack Hall's work and his relations with his wife and son. As the awards for visual effects and sound attest, the special effects are well done, though embellished. The film contains scientific exposition as well, though again, with an unrealistic tempo of events. Nevertheless, Jack's speeches to the U.N. delegates and to the president include true scientific nuggets about global warming. In this way, *The Day After Tomorrow* draws attention to a real and current problem that has a better chance of doing serious harm than any asteroid strike or alien invasion.

The Day After Tomorrow wasn't made as a public service: it was made to earn money by entertaining people. So were *On the Beach*, *The Day the Earth Stood Still*, *Fat Man and Little Boy*, *The China Syndrome*, *The Sixth Day*, *Jurassic Park*, *Panic in the Streets*, *The Insider*, *Outbreak*, and *The Island*. Still, these films, and others, relate problems involving science as cause or cure and describe how science interacts with government and the private sector over issues of nuclear catastrophe, genetic engineering, and public health.

Now real-world science is discerning the threat of global warming and showing how to attack it by reducing carbon dioxide emissions from human activities. Pollution and global warming are the back story in *Soylent Green*, *Waterworld*, and *A.I.*, but *The Day After Tomorrow* spells out the science and the government's reaction. In the movie, Jack works for NOAA, the National

Oceanic and Atmospheric Administration. This real government agency carries out research in climate, weather, the oceans, and the environment, under the motto "Science, Service, Stewardship." Jack's efforts to live by the motto and present his findings to his own employer are rebuffed by the vice president, who dismisses Jack's data because of its impact on industry.

This fictional situation is uncomfortably close to present attitudes, where, unlike other governments, the United States has for some time rejected evidence for global warming and the need to reduce greenhouse gases. So despite its imperfections, *The Day After Tomorrow* deserves credit for displaying some of the real science and for illustrating the conflict that can and does occur when scientific findings clash with government policies or political agendas.

Golden Turkeys

In contrast to the Golden Eagles, films that say something meaningful about science and scientists or are relevant to how we live, there are the Golden Turkeys, films that fail in cinematic or scientific terms, or both. As *The Golden Turkey Awards* and the list from the Bad Cinema Society show, many science-based films fit these criteria, but here are two of the worst.

The Core (2003). *The Core* is truly remarkable: it features five scientists yet still manages to pack record-setting amounts of scientific misinformation into a short time. It's a toss-up who knows less about science, the script writers or the scientists, but any scientist who thinks the Earth has an electromagnetic field, who believes we can drill through thousands of miles of rock and magma with nary a hitch, or who wants to set off hydrogen bombs inside our planet needs to hand in his lab coat.

For students studying under Josh Keyes, the lead scientist, here's a tip: pretty much everything he says and does in the classroom is nonsense, so be wary of any Ph.D.'s he's handing out. But then, this is a film where vastly experienced Dr. Conrad Zimsky, the scientist the government swears by, says, "That's all science is, is best guess." Possibly Zimsky's cavalier attitude explains the looniness of the science in this film.

Whatever the reason, it's hard to forgive scenes like the one where laser expert Ed "Braz" Brazzleton steps into the Earth's interior, at a temperature

The Core (2003). Microwaves from space, no longer deflected by the Earth's missing magnetic field, take out the Golden Gate Bridge. Fortunately, nearly every bit of science in this scene (and in the entire film) is wrong; otherwise, you could never again dare make microwave popcorn. *A Golden Turkey.*

Source: Horsepower Films/The Kobal Collection.

of 9,000 degrees Fahrenheit, and pants a bit but miraculously lasts long enough to save the mission, then dies a clichéd hero's death. In fact, all but two of the intrepid crew that drills down to the Earth's core dies, and pseudo-tragically—meaning that the film never gets us sufficiently connected with these people to care about them. Any emotional contact is shattered by phony lines like this one from weapons expert Serge Leveque: "I came here to save my wife and my two children and . . . six billion lives. . . . It's too much. I just hope I'm, I'm smart enough and brave enough to save three."

Having failed in science, characterization, and dialogue, *The Core* has one more shot at salvation: deep inside the Earth, you might expect to see some amazing sights. But alas, the film's special effects are nothing special, and scenes with giant diamonds, molten magma, and H-bomb explosions fall equally flat. *The Core* stands as *the* prime example of a science fiction film where neither the human side nor the visual impact grabs you and where the science isn't just a little bit wrong: it's so wrong that willing suspension of disbelief is replaced by gales of laughter.

What the #$!Do We (K)now!?* or *What the Bleep Do We Know!?* (2004). At least *The Core* doesn't serve any particular agenda or pretend to be more than what it is: a science fiction story. *What the Bleep* is different and insidious: it's presented as part documentary and part story, combining fictional elements with what seems to be contemplation of deep scientific issues. But what isn't revealed up front is that it represents a view that comes from a group with peculiar ideas that are roundly rejected by real scientists.

The fictional part is the story of Amanda (Marlee Matlin), a photographer. Amanda is recently divorced, and unhappy about that and other aspects of her life, including her physical appearance. Her angst is apparent as she wanders through the city and interacts with others. Interspersed with this saga are talking head shots of about a dozen serious-looking men and women, each speaking at length and with great earnestness. Eventually, we realize that these sages are focusing in different ways on one key idea, "shaping one's own reality."

What they mean by this is a mangled version of quantum physics, the theory of the smallest things in the universe like electrons, photons, and quarks. Quantum theory began around 1900; more than a century later, it has proven essential for us to understand the universe and manipulate our world. It works on a theoretical level, and a practical one too: large chunks of the technology we use daily, from MP3 players, computers, cell phones, and the Internet to lasers used in eye surgery and in supermarket checkout lines, trace back to quantum ideas.

But there are puzzles built into the theory, including the Uncertainty Principle, first stated by the German physicist Werner Heisenberg in 1927. This says that you can never know all there is to know about the behavior of tiny bits of matter and energy such as electrons. Physicists struggle with this idea and with the randomness it implies at the heart of reality, and Einstein didn't like it at all—it led to his famous statement that "God doesn't play dice" with the universe. Nevertheless, this seems to be how the world works. And then there's the role of the observer. In quantum theory, when you observe a physical event by doing an experiment, that act selects one out of a menu of many possibilities, but—and this is crucial—only for tiny components of the universe like electrons.

The Uncertainty Principle and the role of the observer have meaning only at a sub-sub-microscopic level, not for ordinary existence, where events do happen in predictable ways with or without an observer. *What the Bleep* makes a giant, unjustifiable leap when it concludes that we can shape—literally physically sculpt and resculpt—the human-scale world by thought alone. Once this message becomes clear to Amanda, she finds she can change her body and slim her thighs just by wishing it, and the film ends on this happy and uplifting note.

These strange ideas don't come from physics but out of the heads of those who made the film and comprise some of the interviewees, adherents of Ramtha's School of Enlightenment in Yelm, Washington. According to J. Z. Knight, founder of the school, Ramtha is a 35,000-year-old warrior and ascended master from the ancient continent of Lemuria, now sunk beneath the waves. She should know: she channels Ramtha and appears as one of the film's gurus.

Other followers of Ramtha present their ideas in the film, such as chiropractor Joe Dispenza (who, however, later severed his connection with the

What the #$!Do We (K)now!?* (alternate title: *What the Bleep Do We Know!?*) (2004). Unhappy photographer Amanda (Marlee Matlin, left) finally smiles as she begins to discover that according to the rules of quantum mechanics, she can change reality and even firm up her thighs through sheer thought power. Sadly, though, the film completely misinterprets what quantum physics says. *A Golden Turkey.*

Source: Lord of the Wind/The Kobal Collection.

School of Enlightenment). Also prominently shown is the research of Dr. Masaru Emoto, who photographs "water crystals" (these closely resemble ice or snow crystals) as they respond to labels placed on their containers, prayer, and spoken words; for instance, they show different shapes for the names "Adolf Hitler" and "Mother Teresa."

Among these pontificating sages, channelers, and chiropractors, one interviewee with believable scientific credentials is David Albert, a philosopher of physics at Columbia University. But he was not given a chance to express his ideas, which are contrary to those in the film. As he later complained bitterly in an article in *Salon:*

> I was edited in such a way as to completely suppress my actual views. . . . I am, indeed, profoundly unsympathetic to attempts at linking quantum mechanics with consciousness. . . . I explained all that . . . to the producers of the film. . . . Had I known that I would have been so radically misrepresented in the movie, I would certainly not have agreed to be filmed.

In a review published in *The Guardian*, critic Philip French called *What the Bleep* a "near-demented combination of quantum physics [and] New Age mysticism." The ideas may seem demented, but the film's structure isn't; the careful way in which it pretends to a balanced documentary tone while hiding its true colors and the editing of Albert's remarks suggest a deliberate attempt to mislead. Among all these films, this is the only one that earns the lowest rating of all, "dishonest," making it the biggest Golden Turkey of Science.

Like any kind of film, science-based and science fiction films cover a broad range, from schlocky, uncaring work to compelling dramas of people and ideas. Science can appear in a meaningful way even in science fiction, and even a supposedly serious science film can distort and mislead. Both possibilities are important because films are intimately bound up in society, reflecting people's attitudes toward science and, at the same time, shaping those attitudes. That leads to my final question: are the movies, especially science fiction films, good or bad for science, scientists, and society in general?

Chapter 10

Hollywood Science vs. Real Science

> The continuing dance between science and science fiction—in which the science stimulates the fiction, and the fiction stimulates a new generation of scientists, [is] a process benefiting both genres.
>
> —Carl Sagan, *Pale Blue Dot* (1997)

> I would like to thank [the] science fiction writers and moviemakers who inspire us, even when they are dead wrong, but especially when they challenge the human spirit to soar beyond itself.
>
> —Rodney Brooks, *Flesh and Machines* (2002)

Watch *The Core*, or go to the Web site "Insultingly Stupid Movie Physics," and you might conclude that Hollywood science is inevitably garbled or mangled, that films promote only bad science and scientific misinformation. For sure, there's no lack of scientific errors in the movies, as "Insultingly Stupid Movie Physics" gleefully points out and as I've noted often enough in this book. Fortunately, as a counterbalance, there are serious science documentaries, like those shown in the highly regarded, award-winning *Nova* series seen on PBS. Besides, some science fiction films present science pretty well, and even those that don't can still have positive effects because movies influence society in different ways. Although getting the science right matters a great deal, and scientists gnash their teeth when science is presented badly, that isn't always the only consideration—even sometimes for the scientists themselves.

When Cynthia Breazeal, who developed the emotionally responsive robot "Kismet," was interviewed by the *New York Times* in 2003, here's how she answered a key question:

Q. What is the root of your passion for robots?

A. For me, as for many of us who do robotics, I think it is science fiction. My most memorable science fiction experience was "Star Wars" and seeing R2D2 and C3PO. I fell in love with those robots.

Breazeal turned that childhood fascination into a scientific career. Though robots like R2D2 and C3PO were only fantasies when *Star Wars* was released in 1977—and still are—Breazeal's work is a step toward their realization. NASA's Donna Shirley also journeyed from imagination to a reality that she helped create. She tells how, as a child, she was enthralled by science fiction images of Mars. That was a perfect beginning for a woman who earned an aerospace degree and then virtually landed on Mars; she managed the team responsible for the *Sojourner* rover that explored the planet in 1997, and went on to run NASA's Mars Exploration Program.

Such anecdotes from researchers show how science fiction generated a sense of wonder that enhanced their youthful interest in science. For them, it didn't much matter whether the fictional science was exactly right. In fact, these protoscientists were stimulated and challenged by imagined science and technology that wasn't a reality—yet. That forward-looking aspect is science fiction's most valuable property. As population biologist Robert May, former president of the Royal Society, has put it, "At its best [science fiction] is very provocative and forethoughtful, like [Isaac] Asimov's books—they think of questions that have since arisen such as intelligent machines."

The connections between science and science fiction form an ongoing, mutually beneficial relationship. Film director and writer James Cameron, who sits on the Advisory Board of the Science Fiction Museum in Seattle, comments that "science fiction inspire[s] people to become scientists and want to ask questions about the real nature of existence and matter and reality . . . what they're finding then feeds back into the science fiction community . . . and spins out a whole new generation of science fiction."

In his book *The Seven Secrets of How to Think Like a Rocket Scientist*, Purdue astronautics professor James Longuski describes the rocket scientists at NASA's Jet Propulsion Laboratory who gathered regularly to watch science fiction classics like *The Day the Earth Stood Still* and turkeys like *Plan 9 From Outer Space*. They'd laugh at the scientific errors, but "they loved these films.

They were like children who want to hear the same fairy tale over and over again. These were the fairy tales of the rocket scientists; their unfettered hearts seeking contact with outer space. Their logic turned off . . . their dreams turned on. Imagination wasn't silly to them."

Certainly there's active feedback between real science and Hollywood science. The science of the time inspires science on-screen, from nuclear weapons in the 1950s to genetic manipulation in the 1990s. There's an opposite flow as well. The Web site Technovelgy.com ("where science meets fiction") lists over 1,000 inventions and ideas predicted in science fiction stories. Many have come to pass, not because science fiction writers are clairvoyant but because they extrapolate the scientific and social currents of their time. In 1944, well before the atomic bomb was built and tested, science fiction author Cleve Cartmill presented the idea of a nuclear bomb driven by a chain reaction in his story "Deadline." Since the Manhattan Project was top secret, the War Department investigated. Cartmill was cleared of charges of revealing classified data when it became apparent that the information was openly available. A person knowledgeable about physics could work out how to construct a bomb, once he knew about the discovery of nuclear fission in 1938.

The most famous carryover from a science fiction film to the real world may be the invention of the backward countdown, "ten . . . nine . . . eight . . ." before a spacecraft blasts off, as used by NASA. This dramatic overture to a launch first appeared in Fritz Lang's 1929 silent movie *Frau im Mond* (*Woman in the Moon*). In general, though, interactions between science and science fiction involve books as well as movies. Books are better than films at conveying complex ideas (although documentary films can do an excellent job). Classic science fiction authors, such as Isaac Asimov, Arthur C. Clarke, and Robert Heinlein, and contemporary writers, such as Neal Stephenson and Kim Stanley Robinson, have inspired and educated many.

But the enormous power of films to reach millions can't be downplayed, and it's not as if literary and cinematic science fiction are at odds. Although the written form came first, books and films converged back in 1902, when Jules Verne's idea of a cannon shot to the Moon appeared in Georges Méliès *Voyage to the Moon*. Since then, film after science fiction film has come from published works. The many cinematic spinoffs from Mary Shelley's *Frankenstein* are the best known examples, but other deep-literary antecedents abound. For instance, *The Thing from Another World* and its 1982 remake *The*

Thing are based on the 1938 John W. Campbell story "Who Goes There?" This in turn has similarities in its setting and vegetable alien to the 1936 story *At the Mountains of Madness* by the distinguished American horror and science fiction author H. P. Lovecraft. Many films from literary sources have been among the most significant in the genre. Eight of the Golden Eagles in chapter 9 come from books or short stories, as do four of the eight science fiction films on the American Film Institute's list of the hundred best American films.

These dual manifestations aren't redundant, because films and books enhance each other. As Susan Sontag has said in her essay "The Imagination of Disaster," each medium has its own strengths: science fiction books give an "intellectual workout," and films provide "sensuous elaboration." The distinguished film critic Stanley Kauffman puts it this way: "Film floods into a viewer's system of responses much more engulfingly than a book can. In a film, most of the work of transmutation from words to effect has already been done for the viewer." If books excel in conveying scientific ideas, films are wonderful in generating the look, feel, and human impact of a scientific event or a projected future. *On the Beach* and *Gattaca*, for example, each present strong images of future worlds formed by science and its outcomes, with almost no exposition.

But this isn't to say that science fiction films can't also educate. That's good; we need every bit of science teaching we can muster for the general public and for students from elementary school to college. *Science and Engineering Indicators 2006*, the latest report on U.S. science and technology from the National Science Board, shows that our science literacy is nothing to crow about. Stop one hundred adults on the street, and ask a basic scientific question, such as:

- Does the Earth go around the Sun or the Sun around the Earth?
- True or false, antibiotics kill viruses as well as bacteria
- True or false, electrons are smaller than atoms
- Tell me in your own words, what is DNA?

Typically only fifty or sixty of that hundred will know the right answers (71 percent of adults know that the Earth goes around the sun, 54 percent that antibiotics cannot kill viruses, 45 percent that electrons are smaller than atoms, and only 33 percent could define DNA). Equally troubling, surveys show that nearly one-third of American adults believe astrology and fortune telling are "very scientific" or "sort of scientific." These findings relate to concerns about the quality of U.S. science education, which is not uniformly high. Our elementary school students score above international averages in mathematics and science, but high school students score below average. Only one-third of fourth and eighth grade students, and less than one-fifth of twelfth grade students, reach scientific and mathematical proficiency at their grade levels.

Observers worry that poor science education will lead to decreased support for science and fewer students seeking scientific careers; in fact, the current U.S. production of college graduates in science and engineering trails the growth of jobs in these areas, with the difference made up by people from abroad. Most important, the failure to convey science to the populace means we lack a well-informed U.S. citizenry when it comes to the scientific issues confronting us all, from stem-cell research to hurricane prediction and nuclear power.

Serious science education is the job of schools and universities, but people also learn science from other sources. According to *Science and Engineering Indicators 2006*, the main source of scientific information for the American public is television network news and network magazine shows like *60 Minutes*, with the Internet coming up rapidly. People also hear about science from other television and radio programs and books, magazines, and newspapers. Among this array of sources, films play a role in both informal and formal education.

Documentary films that use the power of the visual to present science can be seen on television, but rarely on the biggest broadcast channels; rather, they appear on Public Broadcasting outlets and on cable and satellite outlets such as the Discovery and National Geographic channels, with substantial but smaller viewerships. Among Hollywood feature films widely distributed in theaters and eventually on DVD and television, biopics about scientists also convey something about their science as well as their lives, as in the illustration of John Nash's breakthrough mathematical theorem in *A Beautiful Mind*. Education, however, isn't limited to these genres; science fiction feature films can teach large numbers of people, although this process needs care, caution, and guidance.

At one level, these films educate informally by presenting ideas like emerging viruses and black holes. But when the filmmaker's first commitment is to tell an entertaining story, there's little guarantee that the science is accurate. That may not affect the inspirational value of these films, but mistakes or deliberate distortion can do more harm than good to science literacy and public understanding of science.

Nevertheless, there often is real scientific input into science fiction films in addition to what the screenwriters provide. According to Scott Frank, an anthropologist who studies the culture of Hollywood, filmmakers have good reasons to get the science right. Some directors and producers like scientific themes and are committed to presenting them realistically. Some seek verisimilitude in their movies as a matter of artistic and professional integrity. Actors are likely to have similar motivations; they want their portrayals to be convincing. The most important reason, though, according to Frank, is that today's sophisticated movie audiences want to see scientific realism on screen, and therefore the movie industry believes that scientific realism pays off at the box office.

To inject that realism, filmmakers hire consultants to monitor the science in a film. Frank calls such consultants "mediators between dramatic and veritable truth" and lists many popular science fiction movies that have used them, such as the *Jurassic Park* series, *Twister*, *Volcano*, *Dante's Peak*, *Gattaca*, *The Core*, *Armageddon*, and *Deep Impact*. "At this point," he comments,

> it is almost inconceivable to have a major motion picture . . . that features scientists or scientific themes without hiring a consultant, if purely for PR purposes (as some consultants, like the one who worked on *Volcano*, appear to be). At the dawn of the twenty-first century, even films that seem to have at most a tenuous relationship to "real" science, like *Spider-Man*, have science consultants.

Many consultants bring excellent credentials and some films use them heavily. According to the Internet Movie Database, for instance, *Contact* employed four experts with backgrounds in radio astronomy, NASA operations, and mathematics, with additional advice from other scientists. The result is the sense of scientific verisimilitude the film projects. *Deep Impact* and *Armageddon* each list several consultants (including one they share), with expertise in space technology (two are former astronauts), astronomy, asteroids, and comets. Several geoscientists were consulted for *The Core*, making its director,

Jon Amiel, confident about its science. In publicity for the movie, he said, "Basically, the film is science *faction*: a good dollop of science, a considerable amount of fact and a wee bit of fiction!"

But as I'll show later, there's a blooper in *Armageddon* despite its use of consultants, and I've pointed out the problems with *The Core*. Director Amiel's heartening estimate of the ratio of fact to fiction became mysteriously and laughably inverted in the final result (which bombed at the box office, apparently illustrating that audiences do want a degree of verisimilitude: people can suspend only so much disbelief). So even if consultants aren't brought in strictly for publicity value, hiring them still does not guarantee accuracy. Often it's enough that a film seem "sciencey" to convince audiences, meaning that consultants may advise more about the look of a film than about its science. And when a consultant does want to correct the actual science, the director or producer may not accept the change for economic reasons or because it's no easy feat to balance veritable truth against dramatic truth: too much real science, and you get a flat story like *Destination Moon*; too little, and you get a mess like *The Core*.

Although the science in science fiction films isn't necessarily reliable, these films can still help in formal science education. One pioneering effort has come from physicist Leroy Dubeck at Temple University, who, with his colleagues, began writing about science education through films in 1988. More recently, in 2002, Costas Efthimiou and Ralph Llewellyn at the University of Central Florida introduced a course called Physics in Films. For example, they use *Armageddon* to present ideas of mechanics. As you'll recall, in the story a team of oil drillers implants an H-bomb into an asteroid hurtling toward our planet. The idea is that the nuclear blast will split the asteroid in two and give each piece a sideways push that will make it miss the Earth.

Students in Physics in Films are asked to combine data in the film, such as the fact that the asteroid is the size of Texas, with basic ideas of mechanics, such as momentum and kinetic energy, to calculate the separation between the two halves. They find that the nuclear explosion does indeed push the two halves apart, but not enough to miss the Earth. As Efthimiou and Llewellyn write,

> The students are astonished. Instead of being hit by one Texas-size asteroid, Earth will be hit by two half-Texas-size asteroids

> about 400 meters [1,200 feet] apart! This discussion concludes with an explanation of what is realistically possible and why the government has an ongoing project to detect and track space objects approaching Earth.

Although the plan in *Armageddon* wouldn't work, the basic science is valid; that is, physics predicts that the explosion could split the asteroid and would push the halves apart, just not far enough. This shows how, in the right hands, correcting incorrect science can teach scientific principles. According to data gathered by Efthimiou and Llewellyn, the use of movies keeps students more interested than when problems are expressed in dry or abstract terms, and students in Physics in Films do better than those in a nearly identical course without the films.

But how about a film like *The Core*, which moves toward pseudoscience, that is, unverified claims or intentionally distorted facts presented as mainstream science. The *Virgil*, the craft that reaches the core, is made of a material that gets stronger as it gets hotter and also turns heat into useful energy—properties that contradict known physical behavior and laws. Yet even distorted science like this can be used to teach real science and, equally important, to develop the critical ability to separate fact from fiction. Efthimiou and Llewellyn include a section on pseudoscience in their course, drawing, for instance, on *Independence Day* to discuss extrasensory perception. Even *What the Bleep*, which goes far astray in its version of valid science, provides a perfect opportunity to explain what quantum mechanics really does and doesn't say.

An ambitious program called CISCI (Cinema and Science) that follows these principles is now underway. Funded by the European Union, CISCI involves nine European nations and the United States. European educators and scientists note the same trend seen in the United States, that "science is a subject which is declining and at best stagnating" in popularity among the young, leading to an "alarming" shortage of scientists. To combat this, CISCI aims to

> combine the two most popular media among youngsters, namely movies and the Internet . . . to stimulate interest in science while dispelling widely-spread misconceptions that arise from pseudo-science. CISCI is setting up a free database with

> video clips and movie scenes taken from documentaries and popular movies that . . . illustrate scientific concepts.

The scientific content is monitored by university scientists, and an example of their analysis for *Deep Impact* can be found online (www.scienceinschool.org/repository/docs/deepimpact.pdf). The CISCI database is also available on the Web for school teachers and their pupils and for the general public. One issue, however, is acquiring intellectual property rights for the films.

Another project also shows the power of visual media to convey science, although this effort uses television rather than film. The initiative, called "We All Use Math Every Day" (http://www.weallusematheveryday.com/tools/waumed/home.htm) began in 2005 and is tied in to the weekly CBS TV show *NUMB3RS*. In the continuing story, FBI agent Don Eppes (Rob Morrow) draws on the expertise of his mathematician younger brother Charlie (David Krumholtz) to solve baffling cases. The mathematics is reviewed by professional mathematicians and is well integrated into the stories. The show has proven so popular that CBS, Texas Instruments, and the National Council of Teachers of Mathematics are collaborating to turn it into a teaching resource. "We All Use Math Every Day" provides teachers with classroom activities tied to the mathematics featured each week, which students (and their parents!) can view in the appropriate episode.

These educational efforts focus on science content, but maybe more important is conveying a sense of big issues involving science that affect humanity. When new, little-understood possibilities and threats appear, science fiction films can inform, predict, and warn. Science fiction author David Brin takes it even further. "The most powerful form of science fiction," he said in a recent interview

> is the self-preventing prophesy; not something that predicts the future but something that prevents a future; the best examples being George Orwell's "1984," the movie "Soylent Green" that recruited millions of new environmentalists to help try to save the world, [and] "Dr. Strangelove," which might have helped us to prevent accidental nuclear war.

It's claiming too much to assert that Orwell's book and the two films actually prevented the dire events they predict, or that *Soylent Green* recruited "millions" of environmentalists, but these films certainly helped bring their issues into the popular consciousness. I teach a course at my university with a colleague from film studies that focuses on this aspect; we use science fiction films to present genetic engineering, nuclear power, and other current concerns to students.

One recent effort shows that films indeed affect public perceptions of oncoming risk. Anthony Leiserowitz, an expert in environmental science and policy, has analyzed the response to *The Day After Tomorrow* and its treatment of global warming. This film enjoyed major success. It grossed $69 million in its opening weekend (Memorial Day, 2004), and its worldwide gross to date exceeds $540 million (figures from boxofficemojo.com). Shortly after its release, some 21 million American adults had seen it, with the number climbing after it appeared on DVD.

Leiserowitz surveyed 529 people, a representative sample of U.S. adults, before and after *The Day After Tomorrow* opened in theaters and compared responses from those who had seen the film and from those who had not. He asked about concerns over global warming and the possibility of changing one's own behavior as a result and about political preferences. For those who had seen the film, he concluded that it had a "significant impact" on

> climate change risk perceptions, conceptual models, behavioral intentions, policy priorities. . . . The film led moviegoers to have higher levels of concern and worry about global warming [and] encouraged watchers to engage in personal, political, and social action to address climate change and to elevate global warming as a national priority . . . the movie even appears to have influenced voter preferences.

The film caused no change in overall public attitudes toward global warming because its audience, though huge, is only a fraction of the U.S. adult population. Even enormously successful movies aren't seen by a majority of Americans and so can't immediately swing popular opinion. But the effect of *The Day After Tomorrow* could grow with time, and it had a secondary impact. According to Leiserowitz's analysis, media coverage of the film and its science was ten times greater than was accorded the 2001 report of the U.N. Intergovernmental Panel on Climate Change, which solidified worries about global warming.

Global warming is also the subject of the 2006 documentary *An Inconvenient Truth* (which won an Academy Award for best documentary feature in 2007). This film and the responses to it give a rare opportunity to understand how science and film interact under two different circumstances: when the science comes first and when the story comes first. As documentaries go, *An Inconvenient Truth* generated wide interest even before it earned an Oscar, since it features former vice president Al Gore as narrator and explicator. Still, this attention is only a ripple compared to the impact of *The Day After Tomorrow*, as judged by box office sales (which I use as a rough indicator of the number of people who have seen the film, not as a measure of merit). Although by early 2007 *An Inconvenient Truth* had earned $45 million in global box office sales (making it the third-highest-grossing documentary ever), this is a mere fraction of what *Day After Tomorrow* has grossed worldwide.

You'll recall that *Day After Tomorrow* features dedicated, heroic climatologist Jack Hall, who finds clear evidence that major climate change is on its way. His warnings are unheeded by the authorities. As Jack's predictions come true, the world is afflicted by violent weather phenomena, shown through powerful special effects. Even worse, a new ice age descends, and many people die. Finally the U.S. government listens to Jack and saves lives by evacuating people to Mexico. Meanwhile, Jack undertakes a grueling trek to save his son Sam from freezing in New York, but not before Sam has had to fight off marauding wolves. As the film ends, the violent weather ceases and a bright new day dawns.

An Inconvenient Truth has no wolves, no doom-laden background music, and no happy ending. Mostly we see a somewhat portly Al Gore addressing an audience from a stage as he presents the evidence for global warming. *Truth* couldn't be more different from *Day After Tomorrow*, as Gore musters relatively dry graphs, numbers, and research citations to show that the warming trend is real. Although Gore himself is not a scientist, the film is designed to report actual scientific results and to warn about terrible consequences if the science isn't taken seriously.

But look more closely at both movies and you can see how science and filmmaking bleed into each other. In *Truth*, Gore deploys state-of-the-art multimedia displays, far fancier than in the average lecture, to enliven his fact-based talk, along with strong visuals from photos and film clips. Also, snippets about the college experience that awakened Gore's interest in global

An Inconvenient Truth (2006: Academy Award, best documentary feature). Former vice president Al Gore uses images, along with graphs and numbers, to present the scientific evidence for global warming. Not every scientist agrees with all of Gore's projections, but there is broad consensus about their general trend. The film makes it possible to check the science behind its conclusions and so provides some of the real science that a feature film like *The Day After Tomorrow* cannot.

Source: Lawrence Bender Prods./The Kobal Collection/Eric Lee.

warming, and about other moments in his life, add a touch of personal story to the diet of straight information. And although there is broad scientific consensus for the big picture Gore paints, some climate scientists still think that the film exaggerates and overly dramatizes the perils and the speed at which they would happen.

Conversely, among stunning special effects and Jack's adventures, *Day After Tomorrow* slips in some straight science as Jack earnestly lectures about the coming ice age. In fact, Al Gore and Jack Hall use the same diagram to give the same explanation of how warming could disrupt ocean currents to reduce temperatures, and both describe how a similar event is thought to have happened thousands of years ago. To dramatize global warming, *Truth* shows Antarctic ice cracking off in huge pieces; to personalize the drama, *Day After*

Tomorrow shows Jack saving his precious data by leaping across a crevasse that suddenly opens, also as Antarctic ice cracks off in huge pieces.

Neither film has the perfect mix of good story and good science, but a double bill of the two, along with additional clarification by a scientist or two, would offer both facts and feelings, a powerful intellectual and sensuous introduction to global warming and its consequences. This illustrates again how science fiction films can contribute to public understanding of scientific issues, if the fictional part is placed in context by appropriate commentary and comparison to factual sources.

If filmmakers are correct in believing that science is a hot topic and that movie audiences want to see at least a semblance of real science, Hollywood ought to be fertile ground for entertaining, money-making films that also do right by science and scientists and even contribute to science education. Certainly, big science fiction films will keep coming. As of this writing, IMDb lists over two dozen in various stages from "announced" to "completed," and slated for release from 2007 to mid-2009. These cover alien invasion and alien disease, time travel, cyborgs and robots, biological warfare, mutation, and other themes both classic and new.

The release of *Spider-Man 3* in spring 2007 shows the continuing blockbuster power of science fiction films. The movie broke all box office records on its opening day and weekend, and within a week had earned a gross worldwide revenue of more than $400 million, on a budget officially stated as $258 million. (Although the $258 million figure places *Spider-Man 3* among the most expensive films ever made, industry rumor has it that its true cost, including marketing and promotion, approaches $500 million, making it far and away the most costly movie ever. Even so, the film is expected to return a handsome profit). Other prominent coming attractions that may reach similar smash hit status include *Jurassic Park IV*; *Fahrenheit 451*, based on the Ray Bradbury story; *Journey 3-D*, based on Jules Verne's *Journey to the Center of the Earth*; and *Avatar* and *Battle Angel*, two efforts from James Cameron.

Maybe some of these films will reach the very best in science fiction, generating a soaring sense of wonder, creating amazing future worlds, provoking creative thought, and bringing us face to face with oncoming problems we need to take seriously, as well as providing gripping entertainment. Why shouldn't they also present good science and realistic scientists? As I've shown, that can be done when the effort is made, and hardly to the detriment of the film.

There's nothing wrong with a well-made science fiction popcorn flick chock full of heroic feats and great special effects. But we can do even better. Maybe the day will come when any film that involves science will be proud to display an industrywide "Good Science Seal" stating that "no scientific concepts were seriously harmed in the making of this film." For movie makers trying to appeal to knowledgeable audiences, this could actually enhance the film's popularity. It would also open the door to educational tie-ins, thus giving the multi-billion-dollar film industry a meaningful stake in the futures of our young people and our society that it now lacks.

Most of all, for us, the millions of movie lovers, it would mean we could sit back and enjoy laser weapons, rocket blastoffs, and fearsome aliens—even the end of the world—to our heart's content, knowing that we're taking in some real science along with thrills and popcorn.

Finding Real Science in the Movies and Beyond

As I argue in this book's last chapter, society and the movie industry could benefit by better presentations of science on screen, but how can this be achieved? What clues can tip off a movie viewer to good science-based films? And how can the coverage of science in a film be extended for those who want to learn more?

If the film comes from a book or original screenplay whose author is knowledgeable about science, that can permeate the movie. Carl Sagan's background as a working astronomer illuminates *Contact*, which Sagan helped write, based on his book of the same name. Science described by someone who lives the subject is more authentic than science described by someone who only writes about it. It's true that not many people combine a scientific background with the ability to write a compelling story, as Sagan did, but there are efforts underway to cultivate the combination. For the last three years, Hollywood's American Film Institute has been educating groups of practicing scientists in the art of writing science-based screenplays. If the writing talent is there, these scientists will produce film scripts with built-in authenticity.

Whoever writes the screenplay, outside science consultants should also be used to check for errors and enlarge the science, but they need to be taken seriously and not chosen merely for cosmetic value. If their input is to be truly useful, they should be brought into the production process sooner rather than later, when change may still be possible without serious disruptions. The right kind of consultant will understand that a film is not Science 101 and can help the filmmaker maintain reasonable scientific veracity while leaving the story undamaged and maybe even enhanced.

Really serious progress, though, would require something beyond a kind of cottage industry, movie-by-movie approach; it would need a way for Hollywood and the scientific community to talk to each other. Scientists desperately want to see Hollywood science (and science in all media) done well. So do their professional organizations like the National Academy of Sciences, the American Association for the Advancement of Science, and the American Physical Society, which speak for large segments of the scientific community. For films, however, it isn't clear if any group can speak for the entire movie industry. Nevertheless, a significant effort by the Alfred P. Sloan Foundation has encouraged the development of a substantial number of science-based films. The foundation presents awards to the best of these and showcases them at significant events such as the Sundance and Tribeca Film Festivals. It has also provided support to a large number of students to write and produce science-based films at leading film schools. And as described in my last chapter, CISCI in Europe and "We All Use Math Every Day" in the United States show that different types of entities can come together to enhance science in the media. Fortunately, there are signs that this could happen for Hollywood movies. One industry group, the Producer's Guild of America, has recently approached the National Academy of Sciences to consider far-reaching ways to improve Hollywood science.

Finally, it's important to appreciate the limitations of what can be achieved on-screen. In a review of a book about the science in *Jurassic Park*, Peter Dodson, professor of anatomy and geology at the University of Pennsylvania, put it this way: "A smashing summer flick by its very nature cannot be a primary source of reliable scientific knowledge. Rather it can provide an invitation to learn." Once the nugget of science is expressed in a film, perhaps inspiring a young person or fascinating an older one, other resources can and should be used to deepen understanding, from popular science books and articles, to Web sites and papers in scientific journals. To show how this can work, I've listed some appropriate sources under "Further Reading and Viewing," for those who want to pursue the science that I've discussed. Of course, these represent only a fraction of what's out there for the interested reader.

Alongside Hollywood Science, There's Popcorn Science

One good thing about watching Hollywood films is that even when the movie doesn't contain much science, our most popular movie snack, popcorn, does. The process that produces that tub of overpriced yet irresistibly delicious fluff involves a good deal of physics, chemistry, and agricultural science. In fact, in 1988, Purdue University awarded one of its favorite alums, Orville Redenbacher of microwave popcorn fame, an honorary doctorate for his decades of popcorn research.

Corn has been in the human diet for millennia, and getting perfectly popped corn has been a human goal for a long time too. Popcorn kernels over 5,000 years old, said to be still poppable, have been found in New Mexico. Centuries before movie theaters and microwave ovens, early American settlers were deep into popcorn. They learned about it from Native Americans and were soon enjoying it with cream and sugar as a breakfast cereal.

The power that transforms a small golden kernel into a white mass forty or fifty times bigger would have been recognized by James Watt, the eighteenth-century developer of the steam engine. Each popcorn kernel contains water that turns into steam when heated, generating considerable internal pressure, over 120 pounds per square inch. This explodes the kernel and pushes out its expanded insides while making a noticeable "pop."

Other types of corn and cereal grains also contain water but don't pop under heat. The secret is the peculiar construction of the popcorn kernel. Its hard and impervious outer shell traps the steam while the pressure builds until there's no holding it back. Inside, the kernel is mostly "carbs"—that is, the carbohydrate called starch. What bursts out after the pop is a complex network of starch bubbles that have been inflated by the steam. Under a microscope, this resembles a foamy construction like the head on a beer or soapsuds filling a sink, which gives popcorn its pleasing texture.

But even with the right kind of kernel, popping isn't guaranteed if the popcorn is too dry or too soggy. The optimum amount of water is 13 to 14 percent by weight. Much of the agricultural science of popcorn is devoted to maintaining exactly that economically profitable level, and to finding new, ever frothier corn hybrids with fewer unpopped kernels.

Besides its scientific depth and its excellent impact on the taste buds, popcorn has had at least one practical application. During World War II, the U.S. government used it as a packing material, since its springy exploded kernels are good shock absorbers (now plastic peanuts are preferred). Unexpectedly, popcorn has a dark side too. Some people employed in the processing of microwave popcorn develop "popcorn worker's lung," a serious respiratory ailment, although this seems to come from the added buttery flavorings, not the popcorn itself. However, there's no reason to think there's any danger from the limited exposure when we microwave our own popcorn or consume it at the movies.

As I've said in this book, a popcorn flick can play a good role in conveying science or inspiring future scientists. If you actually munch some popcorn and reflect on its complexities while watching one of these movies, you'll be taking in even more science by mouth as well as through your eyes and ears.

Further Reading and Viewing

Chapter 1. Looking for Science in the Movies? Check Out Science Fiction Films First

Booker, M. Keith. *Alternate Americas: Science Fiction Film and American Culture.* Westport, Conn.: Praeger, 2006.

Bukatman, Scott. *Blade Runner.* London: BFI Publishing, 1997.

Elsaesser, Thomas. *Metropolis.* London: BFI Publishing, 2000.

Kuhn, Annette, ed. *Alien Zone: Cultural Theory and Contemporary Science Fiction Cinema.* London: Verso, 1990.

——, ed. *Alien Zone II: The Spaces of Science Fiction Cinema.* London: Verso, 1999.

Maxford, Howard. *The A–Z of Science Fiction and Fantasy Films.* London: B. T. Batsford, 1997.

Redmond, Sean, ed. *Liquid Metal.* London: Wallflower, 2003.

Rickman, Gregg, ed. *The Science Fiction Film Reader.* New York: Limelight Editions, 2004.

Sobchack, Vivian Carol. *Screening Space: The American Science Fiction Film.* 2nd ed. New Brunswick, N.J.: Rutgers University Press, 1997.

Sontag, Susan. *The Imagination of Disaster.* London: Vintage, 1994.

Telotte, J. P. *Science Fiction Film.* New York: Cambridge University Press, 2001.

Chapter 2. Alien Encounters

Astrobiology Magazine. www.astrobio.net.

Bohannon, John. "Microbe May Push Photosynthesis Into Deep Water." *Science* 308 (2005): 1855.

DiGregorio, Barry E. "The Dilemma of Mars Sample Return." http://pubs.acs.org/subscribe/journals/ci/31/special/digreg/08digregorio.html.

Grinspoon, David Harry. *Lonely Planets: The Natural Philosophy of Alien Life.* New York: ECCO, 2003.

Minkel, J. R. "All Wet? Astronomers Claim Discovery of Earth-like Planet." *Scientific*

American. http://www.sciam.com/article.cfm?articleID=25A261F0-E7F2-99DF-313249A4883E6A86.

Natural Resources Canada. "Past lives: Chronicles of Canadian Paleontology. The *Hallucigenia* flip." http://gsc.nrcan.gc.ca/paleochron/09_e.php.

Overbye, Dennis. "A Planet Is Too Hot for Life, but Another May Be Just Right." *New York Times*, June 12, 2007.

Parker, Barry R. *Alien Life: The Search for Extraterrestrials and Beyond*. New York: Plenum Trade, 1998.

Pickover, Clifford. *The Science of Aliens*. New York: Basic Books, 1998.

Plaut, Jeffrey J., et al. "Subsurface Radar Sounding of the South Polar Layered Deposits of Mars." *ScienceExpress*, 15 March 2007. http://www.sciencemag.org/cgi/rapidpdf/1139672v1.pdf.

Schilling, Govert. "Habitable, But Not Much Like Home." *Science* 316 (2007): 528.

Schneider, Jean. "The Extrasolar Planets Encyclopaedia." http://exoplanet.eu.

Squyres, S. W. et al. "*In Situ* Evidence for an Ancient Aqueous Environment at Meridiani Planum, Mars." *Science* 306 (2004): 1709–14.

Stover, Dawn. "Creatures of the Thermal Vents." *Popular Science* Special Oceans Issue. http://seawifs.gsfc.nasa.gov/OCEAN_PLANET/HTML/popular_science.html.

Ward, Peter D., and Donald Brownlee. *Rare Earth: Why Complex Life Is Uncommon in the Universe*. Dordrecht: Springer, 2000.

Chapter 3. Devastating Collisions

The Avalon Project. "The Atomic Bombings of Hiroshima and Nagasaki: Chapter 10: Total Casualties." http://www.yale.edu/lawweb/avalon/abomb/mp10.htm.

Barnes-Svarney, Patricia L. *Asteroid: Earth Destroyer or New Frontier?* New York: Basic Books, 2003.

Belton, Michael J. S., et al., eds. *Mitigation of Hazardous Comets and Asteroids*. Cambridge: Cambridge University Press, 2004.

Desonie, Dana. *Cosmic Collisions*. New York: Owl Books, 1996.

Giorgini, J. D. et al. "Asteroid 1950 DA's Encounter with Earth in 2880." *Science* 296 (2002): 132–36.

Marcus, Robert, H. Jay Melosh, and Gareth Collins, eds. "Earth Impact Effects Program." http://www.lpl.arizona.edu/~marcus/crater2.html.

Marusek, James A. "Comet and Asteroid Threat Impact Analysis." 2007 Planetary Defense Conference, March 5–8, 2007, George Washington University, Washington, D.C. http://www.aero.org/conferences/planetarydefense/2007papers/P4-3--Marusek-Paper.pdf.

Paine, Michael, and Benny Peiser. "The Frequency and Predicted Consequences of Cosmic Impacts in the Last 65 Million Years." http://members.optusnet.com.au/mpaineau/paine_bioastronomy02.pdf.

Ward, Steven N. and Erik Asphaug. "Asteroid Impact Tsunami of 2880 March 16." *Geophysics Journal International* 153 (2003): F6–F10.

Weiner, Tim. "Air Force Seeks Bush's Approval for Space Arms." *New York Times*, May 18, 2005.

Chapter 4. Our Violent Planet

"*Dante's Peak*: Reviews and Comments." http://volcano.und.nodak.edu/vwdocs/vw_news/dantespeak.html.

Gore, Albert. *An Inconvenient Truth: The Planetary Emergency of Global Warming and What We Can Do About It*. Emmaus, Penn.: Rodale Press, 2006.

Hansen, James. "Defusing the Global Warming Time Bomb." *Scientific American* (March 2004): 68–77.

Kerr, Richard. "Global Warming May Be Homing in on Atlantic Hurricanes." *Science* 314 (2005): 910–11.

Lay, Thorne, et al. "The Great Sumatra-Andaman Earthquake of 26 December 2004." *Science* 308 (2005): 1127–33.

Masters, Jeffrey M. "*The Day After Tomorrow*: Could It Really Happen?" http://www.wunderground.com/education/thedayafter.asp.

Meehl, Gerald A., et al. "How Much More Global Warming and Sea Level Rise?" *Science* 307 (2005): 1769–72.

National Oceanic and Atmospheric Administration. "Tornadoes . . . Nature's Most Violent Storms." http://www.nssl.noaa.gov/NWSTornado.

Richardson, Anthony J., and David S. Schoeman. "Climate Impact on Plankton Ecosystems in the Northeast Atlantic." *Science* 305 (2004): 1609–12.

U.S. Geological Survey. "Lava-Cooling Operations During the 1973 Eruption of Eldfell Volcano, Heimaey, Vestmannaeyjar, Iceland." http://pubs.usgs.gov/of/1997/of97-724/lavaoperations.html.

Webster, P. J., et al. "Changes in Tropical Cyclone Number, Duration, and Intensity in a Warming Environment." *Science* 309 (2005): 1844–46.

Zebrowski, Ernest. *Perils of a Restless Planet: Scientific Perspectives on Natural Disasters*. Cambridge: Cambridge University Press, 1999.

Chapter 5. Atoms Unleashed

The Avalon Project. "The Atomic Bombings of Hiroshima and Nagasaki: Chapter 10: Total Casualties." http://www.yale.edu/lawweb/avalon/abomb/mp10.htm.

Center for International Security and Cooperation, Stanford University. "Preventing Nuclear Proliferation and Nuclear Terrorism: Essential Steps to Reduce the Availability of Nuclear-Explosive Materials." http://iis-db.stanford.edu/pubs/20855/Prvnt_Nuc_Prlf_and_Nuc_Trror_2005-0407.pdf.

Centers for Disease Control and Prevention. "Acute Radiation Syndrome." http://www.bt.cdc.gov/radiation/ars.asp.

Clery, Daniel. "Fusion Reactor: ITER's $12 Billion Gamble." *Science* 314 (2006): 238–42.

Krock, Lexi, and Rebecca Deusser. "Dirty Bomb: Chronology of Events." Nova Science Programming. http://www.pbs.org/wgbh/nova/dirtybomb/chrono.html.

Rhodes, Richard. *Dark Sun: The Making of the Hydrogen Bomb.* New York: Simon & Schuster, 1995.

——. *The Making of the Atomic Bomb.* New York: Simon & Schuster, 1986.

Seife, Charles. "Outlook for Cold Fusion Is Still Chilly." *Science* 306 (2004): 1873.

U.S. Nuclear Regulatory Commission. "Backgrounder on Chernobyl Nuclear Power Plant Accident." http://www.nrc.gov/reading-rm/doc-collections/fact-sheets/chernobyl-bg.html.

——. "Fact Sheet on the Three Mile Island Accident." http://www.nrc.gov/reading-rm/doc-collections/fact-sheets/3mile-isle.html.

Chapter 6. Genes and Germs Gone Bad

Borio, Luciana, et al. "Hemorrhagic Fever Viruses as Biological Weapons: Medical and Public Health Management." *JAMA* 287 (2002): 2391–2405.

Center for Nonproliferation Studies. "Chemical and Biological Weapons: Possession and Programs Past and Present." http://cns.miis.edu/research/cbw/possess.htm.

Centers for Disease Control and Prevention. "Questions and Answers About Ebola Hemorrhagic Fever." http://www.cdc.gov/ncidod/dvrd/spb/mnpages/dispages/ebola/qa.htm.

DeSalle, Robert, and David Lindley. *The Science of Jurassic Park; or, How to Build a Dinosaur.* New York: Basic Books, 1997.

International Forum for Genetic Engineering. "A History of Genetic Engineering." http://www.ifgene.org/history.htm.

National Human Genome Research Institute. "An Overview of the Human Genome Project." http://www.genome.gov/12011238.

Preston, Richard. *The Hot Zone.* New York: Anchor Books, 1995.

Stokstad, Erik. "*Tyrannosaurus rex* Soft Tissue Raises Tantalizing Prospects." *Science* 307 (2005): 1852.

Yount, Lisa. *Biotechnology and Genetic Engineering.* New York: Facts On File, 2004.

Chapter 7. The Computers Take Over

Brooks, Rodney A. *Flesh and Machines: How Robots Will Change Us.* New York: Pantheon, 2002.

Damasio, Antonio R. *Descartes' Error: Emotion, Reason, and the Human Brain.* New York: HarperCollins, 1994.

Kurzweil, Ray. *The Age of Spiritual Machines: When Computers Exceed Human Intelligence.* New York: Viking, 1999.

Nicolelis, Miguel A. L., and John K. Chapin. "Controlling Robots with the Mind." *Scientific American* (October 2002): 46–53.

Perkins, Sid. “Lamprey Cyborg Sees the Light and Responds.” *Science News* 158 (2000): 309.

Perkowitz, Sidney. *Digital People.* Washington, D.C.: Joseph Henry Press, 2005.

———. “Hi, Robot,” *Atlanta Journal-Constitution*, May 23, 2004.

Telotte, J. P. *Replications: A Robotic History of the Science Fiction Film.* Urbana: University of Illinois Press, 1995.

Turing, A. M. “Computing Machinery and Intelligence.” *Mind* 59 (1950): 433–60.

Chapter 8. Scientists as Heroes, Nerds, and Villains

Cohen, S., et al. “The Autism-Spectrum Quotient (AQ): Evidence from Asperger Syndrome/High-Functioning Autism, Males and Females, Scientists and Mathematicians.” *Journal of Autism and Developmental Disorders* 31 (2001): 5–17.

Flicker, Eva. “Representation of Women Scientists in Feature Films: 1929 to 2003.” *Bridges: The Office of Science and Technology's Publication on Science and Technology Policy.* http://www.ostina.org/html/bridges/article.htm?contribution=154.

Frayling, Christopher. *Mad, Bad, and Dangerous? The Scientist and the Cinema.* London: Reaktion Books, 2005.

Howard Hughes Medical Institute. “Becoming a Scientist.” www.hhmi.org/becoming.

Medawar, P. B. *Advice to a Young Scientist.* New York: HarperCollins, 1981.

Media Resources Center, Moffitt Library, University of California at Berkeley. “Scientists and Medical Doctors in the Movies: A Bibliography of Books and Articles.” http://www.lib.berkeley.edu/MRC/scientistsbib.html.

Milnor, John. “John Nash and *A Beautiful Mind.*” *Notices of the AMS* 45 (1998): 1329–32. http://www.stat.psu.edu/news/conferences/JohnNash/milnor.pdf.

Perkowitz, Sidney. “Female Scientists on the Big Screen.” *The Scientist.* 21 July 2006. http://www.the-scientist.com/news/display/24009.

Steinke, Jocelyn. “Cultural Representations of Gender and Science: Portrayals of Female Scientists and Engineers in Popular Films.” *Science Communication* 27 (2005): 27–63.

Stephan, Paula E., and Sharon G. Levin. *Striking the Mother Lode in Science: The Importance of Age, Place, and Time.* New York: Oxford University Press, 1992.

Traweek, Sharon. *Beamtimes and Lifetimes.* Cambridge, Mass.: Harvard University Press, 1988.

Chapter 9. Solid Science and Quantum Loopiness: Golden Eagles and Golden Turkeys

Academy of Motion Picture Arts and Sciences. “The Official Academy Awards Database.” http://www.oscars.org/awardsdatabase.

The Academy of Science Fiction Fantasy and Horror Films. http://www.saturnawards.org/history_academy.html.

American Film Institute. “AFI's 100 Years . . . 100 Movies.” http://www.afi.com/tvevents/100years/movies.aspx.

Bukatman, Scott. *Blade Runner.* London: British Film Institute, 1997.

Elsaesser, Thomas. *Metropolis.* London: British Film Institute, 2000.

French, Philip. "What The #$*! Do We Know?" *The Guardian.* Sunday, May 22, 2005. http://film.guardian.co.uk/News_Story/Critic_Review/Observer_review/0,,1489418,00.html.

Gorenfeld, John. "Bleep" of Faith. *Salon.com.* September 16, 2004. http://www.salon.com/ent/feature/2004/09/16/bleep/index2.html.

Insultingly Stupid Movie Physics. http://www.intuitor.com/moviephysics.

Lindley, David. *Where Does the Weirdness Go? Why Quantum Mechanics Is Strange, but Not as Strange as You Think.* New York: Basic Books, 1997.

Medved, Harry, and Michael Medved. *The Golden Turkey Awards: Nominees and Winners, the Worst Achievements in Hollywood History.* New York: Putnam, 1980.

Perkowitz, Sidney. "Hollywood Physics," *Physics World* (July 2006): 18–23. http://physicsweb.org/articles/world/19/7/3/1.

Plait, Phil. "Bad Astronomy." http://www.badastronomy.com.

Chapter 10. Hollywood Science vs. Real Science

Broad, William J. "From a Rapt Audience, a Call to Cool the Hype." *New York Times,* March 13, 2007.

CISCI, Cinema and Science. "Launch of a New Innovative On-line Educational Environment for Science." http://www.cisci.net.

Dubeck, Leroy W., Suzanne E. Moshier, and Judith E. Boss. *Fantastic Voyages: Learning Science Through Science Fiction Films.* New York: Springer, 2004.

Efthimiou, Costas J., and Ralph A. Llewellyn. "Avatars of Hollywood in Physical Science." *The Physics Teacher* 44 (January 2006): 28–33.

Frank, Scott. "Reel Reality: Science Consultants in Hollywood." *Science as Culture* 12 (December 2003): 427–69.

Glassy, Mark C. *The Biology of Science Fiction Cinema.* Jefferson, N.C.: McFarland, 2001.

Insultingly Stupid Movie Physics. http://www.intuitor.com/moviephysics.

Leiserowitz, A. "Before and After *The Day After Tomorrow.*" *Environment* 46 (2004): 22–37.

Longuski, James. *The Seven Secrets of How to Think Like a Rocket Scientist.* New York: Springer, 2006.

National Science Board. *Science and Engineering Indicators 2006.* http://www.nsf.gov/statistics/seind06.

Plait, Phil. "Bad Astronomy." http://www.badastronomy.com.

Texas Instruments, National Council of Teachers of Mathematics, and CBS. "We All Use Math Every Day." http://www.weallusematheveryday.com/tools/waumed/home.htm.

Appendix: Alongside Hollywood Science, There's Popcorn Science

Byrne, Gregory. "Purdue Keeps Popping." *Science* 240 (1988): 1411.

McKinley, Jesse. "Flavoring-Factory Illnesses Raise Inquiries." *New York Times*, May 7, 2007.

Perkowitz, Sidney. *Universal Foam: From Cappuccino to the Cosmos*. New York: Walker, 2000.

The Popcorn Board. "Popcorn!" http://www.popcorn.org.

Raloff, Janet. "Breath-Taking Popcorn." *Science News Online* 161 (May 11, 2002). http://se02.xif.com/articles/20020511/food.asp.

Sampson, Mark T. "The Chemistry of Popcorn: It's All About Pop-ability." The American Chemical Society. http://www.chemistry.org/portal/a/c/s/1/feature_ent.html?id=c373e9038cc6643f8f6a17245d830100.

Filmography

This filmography lists each film mentioned in the text, with its director and date of release (generally in the United States) as given by the Internet Movie Data Base (www.imdb.com) and other sources.

A.I.: Artificial Intelligence (Steven Spielberg, 2001)
Alien (Ridley Scott, 1979)
Armageddon (Michael Bay, 1998)
Back to the Future (Robert Zemeckis, 1985)
Battlefield Earth: A Saga of the Year 3000 (Roger Christian, 2000)
A Beautiful Mind (Ron Howard, 2001)
The Beginning or the End (Norman Taurog, 1947)
Blade Runner (Ridley Scott, 1982)
A Boy and His Dog (L. Q. Jones, 1975)
The Boys from Brazil (Franklin J. Schaffner, 1978)
A Brief History of Time (Errol Morris, 1991)
Bringing Up Baby (Howard Hawks, 1938)
Chain Reaction (Andrew Davis, 1996)
The China Syndrome (James Bridges, 1979)
Cinderella (*Cendrillon*) (Georges Méliès, 1899)
A Clockwork Orange (Stanley Kubrick, 1971)
The Clonus Horror (alternate titles: *Parts: The Clonus Horror* or *Clonus*) (Robert S. Fiveson, 1979)
Close Encounters of the Third Kind (Steven Spielberg, 1977)
Code 46 (Michael Winterbottom, 2004)
Colossus: The Forbin Project (Joseph Sargent, 1970)
Contact (Robert Zemeckis, 1997)
The Core (Jon Amiel, 2003)
Critical Mass (Fred Olen Ray, 2000)
Dante's Peak (Roger Donaldson, 1997)
The Day After Tomorrow (Roland Emmerich, 2004)

The Day After Trinity (Jon Else, 1981)
The Day the Earth Stood Still (Robert Wise, 1951)
Deep Impact (Mimi Leder, 1998)
Destination Moon (Irving Pichel, 1950)
Dr. Strangelove, or: How I Learned to Stop Worrying and Love the Bomb (Stanley Kubrick, 1964)
E.T.: The Extra-Terrestrial (Steven Spielberg, 1982)
Fail-Safe (Sidney Lumet, 1964)
Fat Man and Little Boy (Roland Joffé, 1989)
Five (Arch Oboler, 1951)
Frankenstein (James Whale, 1931)
Gattaca (Andrew Niccol, 1997)
Godzilla (*Gojira*) (Ishirô Honda, 1954)
Goldeneye (Martin Campbell, 1995)
Good Will Hunting (Gus Van Sant, 1997)
Gorillas in the Mist: The Story of Dian Fossey (Michael Apted, 1988)
An Inconvenient Truth (Davis Guggenheim, 2006)
Independence Day (Roland Emmerich, 1996)
Infinity (Matthew Broderick, 1996)
The Insider (Michael Mann, 1999)
Invasion of the Body Snatchers (Don Siegel, 1956)
Invasion of the Body Snatchers (Philip Kaufman, 1978)
I.Q. (Fred Schepisi, 1994)
I, Robot (Alex Proyas, 2004)
The Island (Michael Bay, 2005)
The Island of Dr. Moreau (Don Taylor, 1977)
The Island of Dr. Moreau (John Frankenheimer, 1996)
Island of Lost Souls (Erle C. Kenton, 1933)
Jurassic Park (Steven Spielberg, 1993)
Jurassic Park III (Joe Johnston, 2001)
Kinsey (Bill Condon, 2004)
Krakatoa, East of Java (Bernard L. Kowalski, 1969)
Life Story (Mick Jackson, 1987)
The Lost World: Jurassic Park (Steven Spielberg, 1997)
Madame Curie (Mervyn LeRoy, 1943)
Mad Max (George Miller, 1979)
Mad Max Beyond Thunderdome (George Miller and George Ogilvie, 1985)
The Manhattan Project (Marshall Brickman, 1986)
Mars Attacks! (Tim Burton, 1996)
The Matrix (Andy and Larry Wachowski, 1999)
The Matrix Reloaded (Andy and Larry Wachowski, 2003)
The Matrix Revolutions (Andy and Larry Wachowski, 2003)
Meteor (Ronald Neame, 1979)

Metropolis (Fritz Lang, 1927)
Mimic (Guillermo del Toro, 1997)
Night of the Comet (Thom Eberhardt, 1984)
On the Beach (Stanley Kramer, 1959)
On the Beach (Russell Mulcahy, 2000)
Outbreak (Wolfgang Petersen, 1995)
Panic in the Streets (Elia Kazan, 1950)
Plan 9 from Outer Space (Edward D. Wood Jr., 1959)
Proof (John Madden, 2005)
The Puppet Masters (Stuart Orme, 1994)
Real Genius (Martha Coolidge, 1985)
The Road Warrior (*Mad Max 2*) (George Miller, 1981)
RoboCop (Paul Verhoeven, 1987)
Rosalind Franklin: DNA's Dark Lady (Gary Glassman, 2003)
The Saint (Phillip Noyce, 1997)
The Satan Bug (John Sturges, 1965)
Silkwood (Mike Nichols, 1983)
The Sixth Day (Roger Spottiswoode, 2000)
Sky Captain and the World of Tomorrow (Kerry Conran, 2004)
Soylent Green (Richard Fleischer, 1973)
Spider-Man (Sam Raimi, 2002)
Spider-Man 2 (Sam Raimi, 2004)
Spider-Man 3 (Sam Raimi, 2007)
Starship Troopers (Paul Verhoeven, 1997)
Star Trek: First Contact (Jonathan Frakes, 1996)
Star Trek: The Motion Picture (Robert Wise, 1979)
Star Trek IV: The Voyage Home (Leonard Nimoy, 1986)
Star Wars Episode I: The Phantom Menace (George Lucas, 1999)
Star Wars Episode III: Revenge of the Sith (George Lucas, 2005)
Star Wars Episode IV: A New Hope (George Lucas, 1977)
Stealth (Rob Cohen, 2005)
Straw Dogs (Sam Peckinpah, 1971)
The Sum of All Fears (Phil Alden Robinson, 2002)
The Terminator (James Cameron, 1984)
Terminator 2: Judgment Day (James Cameron, 1991)
Terminator 3: Rise of the Machines (Jonathan Mostow, 2003)
Them! (Gordon Douglas, 1954)
The Thing (*John Carpenter's The Thing*) (John Carpenter, 1982)
The Thing from Another World (Christian Nyby, 1951)
Top Gun (Tony Scott, 1986)
Twister (Jan de Bont, 1996)
2001: A Space Odyssey (Stanley Kubrick, 1968)
Volcano (Mick Jackson, 1997)

Voyage to the Moon (*Le Voyage dans la Lune*) (Georges Méliès, 1902)
The War of the Worlds (Byron Haskin, 1953)
War of the Worlds (Steven Spielberg, 2005)
Waterworld (Kevin Reynolds, 1995)
What the #$!Do We (K)now!?* (alternate title: *What the Bleep Do We Know!?*) (William Arntz, Betsy Chasse, and Mark Vicente, 2004)
When Worlds Collide (Rudolph Maté, 1951)
The Wizard of Oz (Victor Fleming, 1939)
Woman in the Moon (*Frau im Mond*) (Fritz Lang, 1929)
The World Is Not Enough (Michael Apted, 1999)

Acknowledgments

As has been true for all my books, I've received wonderful support from my home base, Emory University, including the Film Studies Department and its past and present chairs, David Cook and Matthew Bernstein, respectively. My particular thanks and gratitude go to Eddy von Mueller of that department, my coteacher in our Science in Film course, who used his broad knowledge of films and science fiction to comment incisively on the manuscript, and who has taught me a lot about film.

Other faculty who discussed ideas or commented on the manuscript include Rick Williamon and Phil Segre in physics, Arri Eisen in biology, and Bill Size in environmental studies. Among Emory students, Cary Jones suggested relevant films and tenaciously hunted down references; Richard Wilkes researched some of the science behind the science fiction and carefully fact-checked the manuscript; Daniel Weiss carefully checked film facts; and Erica Sarti and Brittany Kendig, working under Emory's SIRE program, creatively researched the psychology of scientists. In Science in Film, Aileen Carolan, Paul Winterhalter, and other students contributed helpful comments and corrections.

Others helped because they know and love movies, writing, or both. Edna Kleinbaum tracked down scientists in film and the best and worst science fiction movies; Sam Marinov's analysis sharpened my thinking about scientist characters in film; Jake Jacobson carefully reviewed my film facts; and Lois Morris and Bob Lipsyte made creative suggestions. At the U.K. publication *Physics World*, Dens Milne and Matin Durrani greatly assisted in tracking down the images appearing in this book. Lisa Yaszek at Georgia Tech assisted in the publication process. My gratitude goes to all these friends and colleagues for rallying round. Of course, any errors or omissions in this work are my responsibility alone.

My agent, Michelle Tessler, was extraordinarily helpful, especially when publishing details and contracts became unexpectedly complex. My editor, Patrick Fitzgerald, and the staff at Columbia University Press, especially Marina Petrova and Michael Haskell, provided unflagging enthusiasm, bright ideas, and a professional level of support.

My greatest thanks go to my dear wife, Sandy, as always a patient listener and reader and a perceptive editor who has made this a better book; and to my son Mike, who passed on his own suggestions and insights about Hollywood science.

Sidney Perkowitz
Atlanta, Georgia, and Cannon Beach, Oregon
2005–2006

Index

Page locators in italics refer to figures